Springer Proceedings in Complexity

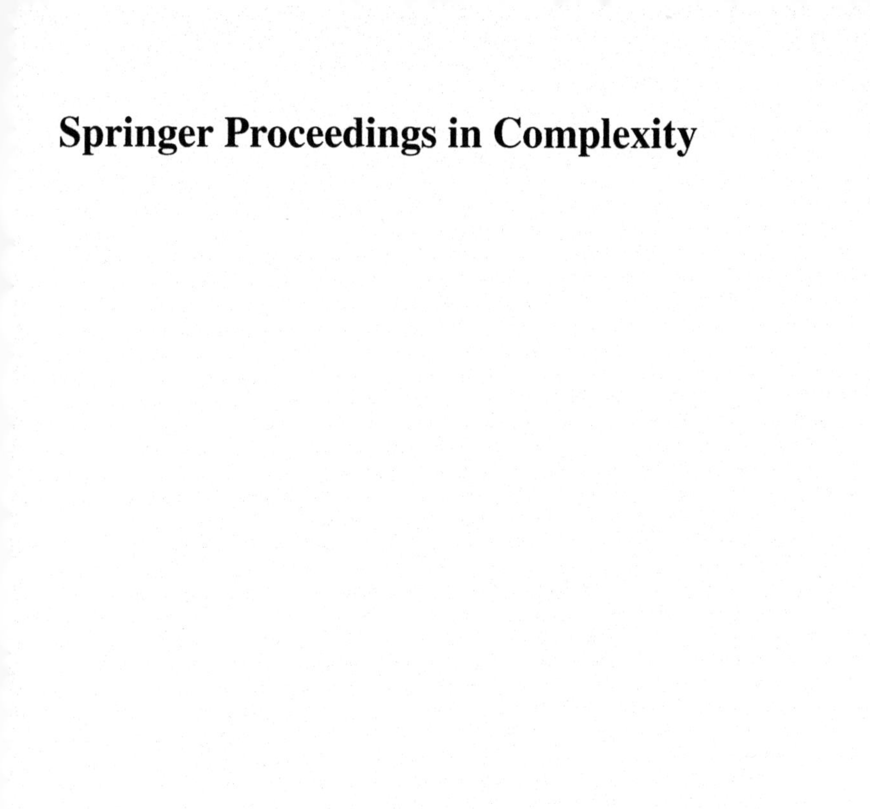

For further volumes:
http://www.springer.com/series/11637

Muhammad Aurangzeb Ahmad • Cuihua Shen
Jaideep Srivastava • Noshir Contractor
Editors

Predicting Real World Behaviors from Virtual World Data

Editors
Muhammad Aurangzeb Ahmad
Dept. Computer Science and Engineering
University of Minnesota
Minneapolis
Minnesota
USA

Jaideep Srivastava
Department of Computer Science
University of Minnesota
Minneapolis
Minnesota
USA

Cuihua Shen
Emerging Media & Communication
Program School of Arts & Humanities
University of Texas at Dallas
Richardson
Texas
USA

Noshir Contractor
Dept. of Communication Studies
Northwestern University
Evanston
Illinois
USA

ISSN 2213-8684 ISSN 2213-8692 (electronic)
ISBN 978-3-319-07141-1 ISBN 978-3-319-07142-8 (eBook)
DOI 10.1007/978-3-319-07142-8
Springer Cham Heidelberg New York Dordrecht London

Library of Congress Control Number: 2014940069

Printed on acid-free paper

Springer is part of Springer Science+Business Media (www.springer.com)

To my father, Mushtaq Ahmad Mirza, you will always be missed; to my mother, Khalida Parveen, may you always be with us

–Muhammad Aurangzeb Ahmad

To my daughter Amelia Lingyun

–Cindy Shen

Preface

In September 2013, I was honored to be the keynote speaker at a workshop at the ASE/IEEE International Conference on Social Computing. The workshop was titled "Predicting Real World Behaviors from Virtual World Data." This book is an outcome of that workshop.

The realm of virtual worlds (VW) and massively multiplayer online games (MMOG) is a fascinating one for social scientists. This online activity has seen extensive growth in the last decade, both in the number of VWs and MMOGs on the market, and the number of players and participants worldwide. The worldwide popularity of this online medium has increased the opportunities for communicating and socializing between individuals and groups.

Researchers showed early enthusiasm for the study of VWs and MMOGs. Scientists postulated that much could be learned about offline behavior by studying behavior online, assuming that the behavior would typically "map" from one environment to the other. The online communities provided a rich natural laboratory where all actions could be captured via the digital footprints left by the participants.

In one example, researchers studied the spread of the "Corrupted Blood" virus in the popular MMOG *World of Warcraft*. The virus was designed by the game developers to spread from one player character to the other, and soon erupted into a full-scale epidemic, with thousands of characters infected in the game.[1] The researchers hypothesized that we could learn much about the spread of viruses in the real world by studying the patterns of infection in virtual environments.

Of course, this "mapping principle"[2] may not apply in all situations. The players in the Corrupted Blood incident may have been trying to deliberately infect one another, just for the fun of it. Then there are experts who postulate that we should not assume too much about the real identity of the individuals that we meet online. People may be exploring alternative personas and ways of behaving in this relatively anonymous setting.[3] The attractive, youthful female avatar in the VW *Second Life*

---

[1] Lofgren, E. T., & Fefferman, N. H. (2007). The untapped potential of virtual game worlds to shed light on real world epidemics. *The Lancet Infectious Diseases, 7,* 625–629.

[2] Williams, D. (2010). The mapping principle, and a research framework for virtual worlds. *Communication Theory, 20,* 451–470.

[3] Turkle, S. (1995). *Life on the screen: Identity in the age of the Internet.* Simon and Schuster.

may be controlled by an older male player. After all, "On the Internet, nobody knows you're a dog".[4]

This view of online behavior suggests that players act out fantasies, deceptions, and exploration of alternate identities. However, maintaining the consistency of this façade requires a large mental effort on the part of the player. When caught up in the thrill of a dungeon raid in *World of Warcaft*, the heat of battle in *Eve Online*, or the pleasure of conversation and music in *Second Life*, it is difficult to maintain a false persona while also concentrating on the task (or play!) at hand. This is simply too much cognitive load for most of us.

As a program manager at IARPA, I had the opportunity to test this in a research program called Reynard. The premise of the program was that our real world (RW) characteristics strongly influence our behaviors, both offline and online, and so those RW characteristics and ways of behaving "bleed over" into VW and MMOG behaviors. Those online behaviors can be digitally captured, analyzed, and modeled to create quantifiable behavioral indicators of RW demographic and personal characteristics of the players. Over a 3-year period, 5 research teams studied over 15,000 players from 9 countries in 12 different MMOGs and VWs. The results of some of this research, along with other related work, are presented in this book.

The "mapping" of the VW and RW can be bi-directional. That is, RW characteristics and experiences can influence an individual's behavior in VWs. Similarly, experiences in VWs can have an impact on behavior and attitudes in the RW. A sampling of the research literature shows that both premises can be supported.

As an example of the RW influencing the VW, studies of various types of social media have found that the offline personality characteristics of the user are portrayed in online behavior. Researchers found they could predict the personality characteristics (as measured by a Big 5 personality scale) of *World of Warcraft* (WOW) players by examining their game play behavior as recorded in the WOW Armory.[5] Behaviors such as questing, playing solo, doing dungeon runs, and completing achievements, were indicative of personality characteristics such as introversion, openness, or conscientiousness.

This premise does not apply just to games. In a study of Twitter, language features were found to be predictive of personality characteristics such as openness and agreeableness.[6] A study of the Chinese messaging service Renren found that type and frequency of blog postings could be predicted from Big 5 personality characteristics.[7]

The research described here has concentrated on predicting a general class of characteristics such as personality, gender, and age. At times, however, an individual may reveal much more about themselves that can lead to specific identification.

---

[4] Steiner, P. L. (1993). http://www.plsteiner.com/ *The New Yorker*.

[5] Yee, N., Ducheneaut, N., Nelson, L., & Likarish, P. (2011). Introverted elves & conscientious gnomes: The expression of personality in World of Warcraft. *CHI-2011*, 753–762.

[6] Golbeck, J., Robles, C., Edmondson, M., & Turner, K. (2011) Predicting personality from Twitter. *Proceedings of the 3rd IEEE International Conference on Social Computing*, Boston, Massachusetts, 149–156.

[7] Bai, S., Zhu, T., & Li Cheng, L. (2012). Big-Five personality prediction based on user behaviors at social network sites.

The choice of character name or screen name is one example. Name choices not only reveal gender and nationality,[8] but also can tie an individual person to multiple accounts. Players reuse the same name, or variations of the name, over and over in multiple games, such as "Gimli", "G!ml!", or "GimLi". This is possibly because (1) the players wish for their friends to be able to find them online in multiple games and social media, and (2) most of the social media sites require unique "handles". Once a user has hit upon a unique combination of letters, numbers, and symbols that no one else has used, they are likely to stick with it.

The influence of the players' experiences in virtual environments can also bleed over into the real world. Researchers at the Stanford University Virtual Reality Lab have found, for example, something that they call the Proteus Effect. By representing the player as a more physically attractive avatar, they were able to increase subsequent positive dating behaviors.[9] Participants in another study watched their avatar exercising online. Those individuals were more likely to report exercising in the RW in the days after the study, as opposed to those whose avatar did not exercise.[10]

The potential to influence RW behavior through the use of VWs and games has not escaped the attention of the medical community. As another example, young cancer patients were randomly assigned to play either a commercial video game (the control condition), or a special videogame called *Remission*, which challenged the young players to destroy cancer cells with a variety of in-game weapons. During the months-long course of the study, researchers found that the participants who played *Remission* had a higher adherence to their treatment protocols than did the control group.[11]

The RW impact of VW activities may not always be beneficial, however. The potential harm from playing videogames has in fact received the majority of the attention from mainstream media. Multiple studies have shown some relationship between playing violent videogames and violent tendencies.[12] This begs the classic scientific question: is this a case of correlation or causation? In those cases where the study has been experimental, with players randomly assigned to play violent versus nonviolent videogames, the effect is still found. If we posit that virtual experiences can have positive effects on RW behaviors, then we must accept that the negative case may also be true.

---

[8] Mateos, P., Longley, P. A., & O'Sullivan, D. (2011). Ethnicity and population structure in personal naming networks. *PLoS ONE 6*(9): e22943.

[9] Yee, N., & Bailenson, J. N. (2007). The proteus effect: The effect of transformed self-representation on behavior. *Human Communication Research, 33,* 271–290.

[10] Fox, J., & Bailenson, J. N. (2009). Virtual self-modeling: The effects of vicarious reinforcement and identification on exercise behaviors. *Media Psychology, 12,* 1–25.

[11] Kato, P. M., Cole, S. W., Bradlyn, A. S., & Pollock, B. H. (2008). A video game improves behavioral outcomes in adolescents and young adults with cancer: A randomized trial. *Pediatrics, 122,* e305–e317.

[12] Anderson, C. A., & Bushman, B. J. (2001). Effects of violent video games on aggressive behavior, aggressive cognition, aggressive affect, physiological arousal and prosocial behavior: A meta-analysis. *Psychological Science, 12,* 353–359.

As a final example of the positive influence of experience in VW environments, consider the large amount of interest in the use of games as training tools. Multiple organizations are experimenting with the use of games for Science, Technology, Engineering, and Math (STEM) education. IARPA, in the Sirius program, is experimenting in this area as well, exploring the use of serious games for teaching individuals to recognize and mitigate cognitive biases in critical thinking.[13] Early research results show that the Sirius games have powerful, long-lasting impacts on judgment and decision-making choices.

As a final thought before you read this book, I will mention another concern that scientists must carefully consider: research ethics. The ability to access large data sets from a variety of social media brings with it a responsibility to the users of that social media. Privacy advocates warn of the possible abuse of large-scale data-mining, and the ability to identify supposedly anonymous individuals through their online activities. Research shows that this identification may indeed be possible.[14]

The question of whether the individuals who have signed up to play MMOGs, or participate in other sorts of social media, have consented to have their captured data used for research, is a vexing one.[15] Most of us bypass reading the Terms of Service, clicking the "I Agree," without understanding that we may have given permission for our online activities to be recorded, sorted through, given to a research team, or sold to marketing companies. Researchers in the area of VWs and MMOGs must continue to hold themselves to the highest ethical standards, complying with applicable human subject research regulations. It is only through the exercise of great care in the protection of human subjects' privacy that research in this area can continue.

Program Manager Rita M. Bush, Ph.D.
Intelligence Advanced Research
Projects Activity (IARPA)
Office of the Director of National Intelligence

[13] Insert URL for Sirius.

[14] Jones, R., Kumar, R., Pang, B., Tomkins, A. (2007). "I know what you did last summer"—Query logs and user privacy. *CIKM'07,* Lisboa, Portugal.

[15] Fairfield, J. A. T. (2012). Avatar experimentation: Human subjects research in virtual worlds, *U.C. Irvine Law Review, 2,* 695–772.

Contents

Contributors

Muhammad Aurangzeb Ahmad University of Minnesota, Minneapolis, MN, USA

Iftekhar Ahmed University of North Texas, Denton, TX, USA

Meg Barton Leidos, Arlington, VA, USA

Jeremy Bernstein Sandia National Labs, Albuquerque, NM, USA

Channing Brown University of Illinois at Urbana Champaign, Urbana, Illinois

Noshir Contractor Northwestern University, Evanston, IL, USA

Geoffrey Cranmer Leidos, Arlington, VA, USA

Komal Kapoor Department of Computer Science and Engineering, University of Minnesota, Minneapolis, MN, USA

Tracy Kennedy Department of Communication, Popular Culture and Film, Brock University, St. Catharines, Canada

Kiran Lakkaraju Sandia National Labs, Albuquerque, NM, USA

Aaron Lawson Speech Technology and Research (STAR) Lab, SRI International, Menlo Park, CA, USA

John Murray Computer Science Laboratory, SRI International, Menlo Park, CA, USA

Amogh Mahapatra University of Minnesota, Minneapolis, MN, USA

Salvatore Marano Dipartimento di Ingegneria Informatica, Modellistica, Elettronica e Sistemistica (DIMES), Università della Calabria, Rende, Italia

Nishith Pathak Department of Computer Science and Engineering, University of Minnesota, Minneapolis, MN, USA

Marshall Scott Poole University of Illinois at Urbana Champaign, Urbana, Illinois

Mary Magee Quinn Leidos, Arlington, VA, USA

Rabindra (Robby) Ratan Department of Telecommunication, Information Studies and Media, Michigan State University, East Lansing, MI, USA

Byron Raines Leidos, Arlington, VA, USA

Floriano De Rango Dipartimento di Ingegneria Informatica, Modellistica, Elettronica e Sistemistica (DIMES), Università della Calabria, Rende, Italia

Jaideep Srivastava University of Minnesota, Minneapolis, MN, USA

Cuihua Shen University of Texas at Dallas, Richardson, TX, USA

Annalisa Socievole Dipartimento di Ingegneria Informatica, Modellistica, Elettronica e Sistemistica (DIMES), Università della Calabria, Rende, Italia

Jaideep Srivastava Department of Computer Science and Engineering, University of Minnesota, Minneapolis, MN, USA

University of Minnesota, Minneapolis, MN, USA

Carl Symborski Leidos, Arlington, VA, USA

Jon Whetzel Sandia National Labs, Albuquerque, NM, USA

Dmitri Williams Annenberg School for Communication and Journalism, University of Southern California, Los Angeles, CA, USA

On the Problem of Predicting Real World Characteristics from Virtual Worlds

Muhammad Aurangzeb Ahmad, Cuihua Shen, Jaideep Srivastava and Noshir Contractor

Abstract Availability of massive amounts of data about the social and behavioral characteristics of a large subset of the population opens up new possibilities that allow researchers to not only observe people's behaviors in a natural, rather than artificial, environment but also conduct predictive modeling of those behaviors and characteristics. Thus an emerging area of study is the prediction of real world characteristics and behaviors of people in the offline or "real" world based on their behaviors in the online virtual worlds. We explore the challenges and opportunities in the emerging field of prediction of real world characteristics based on people's virtual world characteristics, i.e., what are the major paradigms in this field, what are the limitations in current predictive models, limitations in terms of generalizability, etc. Lastly, we also address the future challenges and avenues of research in this area.

1 Introduction

> When the number of factors coming into play in a phenomenological complex is too large scientific method in most cases fails.
>
> —Albert Einstein in *Out of my later years* [12]

Although somewhat simple but it would not be a great exaggeration to state that the Sciences consist of two important parts—the descriptive and the predictive. Yet, for the most part, predicting human behaviors has been a more challenging task than describing them. The reasons behind this are twofold: first, until very recently, it has

M. A. Ahmad (✉) · J. Srivastava
University of Minnesota, Minneapolis, MN, USA
e-mail: mahmad@cs.umn.edu

C. Shen
University of Texas at Dallas, Richardson, TX, USA
e-mail: shencuihua@gmail.com

J. Srivastava
e-mail: srivasta@cs.umn.edu

N. Contractor
Northwestern University, Evanston, IL, USA
e-mail: nosh@northwestern.edu

M. A. Ahmad et al. (eds.), *Predicting Real World Behaviors from Virtual World Data*,
Springer Proceedings in Complexity, DOI 10.1007/978-3-319-07142-8_1,

been extremely difficult and costly to collect accurate behavioral data about human beings in large amounts. Second, human behaviors, the very phenomena that the Social Sciences are studying, are much more complex as compared to the subjects of study of natural sciences. The last 15 years have seen an explosion of digital trace data that can be collected about human behaviors and thus it has also enabled us to not only answer traditional questions in social sciences but also ask new questions, which have not been possible to be addressed because of lack of data in the past. This has given rise to the field of Computational Social Science [19] and researchers in the area have likened the current state of affairs in this field to the circumstances that gave rise to Cognitive Science in the 1950s with the emergence of new types of inter-disciplinary collaborations and availability of new types of data. The emergence of this field offers us new opportunities and challenges with respect to new methods since traditional methods of data gathering and analysis may not scale well or it could be the case that the older methods were not designed with the new type of data in mind.

Massive online games (MOGs) or massively multiplayer online (MMO) games are online games that are characterized by shared persistent online environments where hundreds of thousands, and in some cases, millions of users can simultaneously be part of the virtual environment and interact with one another and also with the environment. Well-known examples of MMOs include World of Warcraft (WoW), EVE Online, EverQuest, EveryQuest II, Star Wars: Knights of the Republic (SWTOR), etc. Given that a large subset of the population in these environments actually spends a significant amount of time in these spaces, the questions arises: do people still have the same persona when they are playing online and how much of their offline persona do they bring online when they are playing in these environments? Hence we would like to know if it is possible to use data from virtual worlds to predict about the "real" world characteristics of players e.g., gender, age, location, deviance, personality, ideology (political), etc. We address these and related questions in this chapter. We consider the application domain and how these virtual worlds map to their real world counterparts as well as where the mapping fails. Additionally, we discuss issues related to generalization of results obtained from studying these environments, how data quality impacts analysis, data augmentation, different approaches that can be used for data modeling, etc. Many of the insights in this chapter are based on our work in the Virtual World Observatory (VWO) project as well as our work in the gaming industry.

2 The Mapping Principle

The Mapping Principle [26] starts with the simple observation that some human behaviors in virtual spaces are often quite similar to their counterparts in the offline flesh and blood world. Dmitri Williams argues that the Mapping Principle cannot be taken for granted in the virtual places, at least not at this point in development of virtual worlds, and has to be established on a case-by-case basis. From this observation it becomes obvious that not all virtual worlds and virtual behaviors map to the offline world. Williams notes that successful mapping in the virtual worlds offers certain advantages, which were not available to research in the social sciences before access to such types of data became available recently. For example, many of the social structures that one observes in the virtual world have offline counterparts, but data collection of such structures in the offline world is often extremely limited by resource constraints. These also offer possibilities with respect to studying human behaviors at different levels of granularity, i.e., at the individual, group, and society levels.

Williams also notes the problem of representation in studying virtual worlds, i.e., different virtual worlds may represent a user in different ways and thus this may make the task of mapping from one virtual environment to another virtual environment quite complicated. This means that we can take it for granted that one virtual world is similar to another virtual world and thus some sort of mapping has to be also established from one virtual world to another world. Even within same type of virtual worlds, different types of social environments can elicit different types of behaviors and should be part and parcel of standard analysis. For example, Player versus Player (PvP) and Player versus Environment (PvE) types of gameplay are quite different with respect to eliciting cooperative or aggressive behaviors from their respective players. Lastly, another complication lies in the fact that even an isolated virtual world is not really isolated since human beings by their very nature bring certain cognitive biases with them. This is clearly evident in the "Proteus Effect" [27] where people's behaviors in the virtual worlds are greatly dictated by the characteristics of their avatars regardless of whether the avatars are similar to their offline characteristics or not.

3 Theory-Driven versus Data-Driven Paradigms

The history of the various sciences is characterized by two different yet complimentary approaches to hypothesis formation and testing: the theory-driven approach and the data-driven approach. We consider these paradigms within the context of social sciences. Given a certain phenomenon, the theory-driven approach takes a social science theory or a set of theories as the guiding principle for data collection and exploration purposes. The type of data that are collected is that guided by or even constrained by theory. Thus consider the phenomenon of friendship and the motivations behind why people form relationships. Social science theories would suggest factors related to homophily, proximity, personality, etc. that would account for the formation of friendship relationships [18, 21]. The data-driven approach, on the other hand, takes a theory-agnostic view with respect to data collection and analysis. While all types of data are collected by considering technological, ethical, and resource constraints within the data-driven paradigm, the analysis is guided by traditional computational concerns like scalability, dimensionality reduction for tractable analysis, information theoretic measures for determining data relevance, etc [15].

It is important to note here that the two approaches are not disjoint; theory is always informed by data. What we mean here by the data-driven approach is that, more often than not, this approach does not presuppose a theory vis-a-vis data collection and its interpretation. Big data add a new dimension to these two approaches. One drawback of the data driven approach is that one can end up with a black box where the variables that one may infer with respect to explaining a social phenomenon may not make sense from a theory perspective. In the ideal case, the mismatch between the theory and the data can reveal new insights with respect to the underlying phenomenon and thus may even help update the theory. In the worst case, these new factors may be treated as epiphenomenon from the perspective of social scientists who want models that are always tied to some social or psychological explanation. In most cases, however, the data-driven approach can indeed be used as a source of feedback to the theory-driven approach where the latter is no longer limited by traditional methods of data collection.

A number of studies have been done in this area to demonstrate the efficacy of the mutually reinforcing nature of these two approaches. Thus, Ahmad et al. [19] addressed the problem of detecting Gold Farmers in MMOs where they used traditional domain rich social science approaches for data analysis and then augmenting that approach with machine learning techniques for feature construction and model building. Similarly Borbora et al. [7] describe the use of both of these approaches for determining churners in MMOs. One set of features are chosen based on literature that describes psychological factors and motivations for engagement and play and another set is chosen based on information theoretic measure like information gain. The main conclusion from their work is that a combination of both these approaches is most accurate for prediction tasks. The final choice of the feature sets can also shed some light on the psychological factors and motivations, which may have been missed by theory.

4 Limitations and Methodological Issues

While Big Data offers us opportunities to gain insights into human psyche and social behaviors, it should not be taken to be the be all and end all of studying human behaviors. There are a number of methodological issues with respect to using log data from virtual worlds and MGOs that preclude one from making certain conclusions about human behaviors. Additionally there are certain limitations, such as data availability, collecting missing data after the fact, issues related to generalization across environments and generalization in the real world, issues related to self-reported data, and how incorrect mapping can actually lead to incorrect conclusions. We discuss each of these issues in some detail in the following subsections and also offer solutions to partially mitigate these problems.

4.1 Data Quality

Game logs and other databases in case of virtual worlds can provide a quite comprehensive view of a user's characteristics and activities within a virtual environment, but they do not always capture everything about the environment being studied, e.g., psychological motivations, reaction to outside events such as divorce, drug abuse, religious conversion, etc. Thus, even in the ideal case where the data logs are recording everything, it is still the case that *everything is not everything*. This particular issue can be partially mitigated by complementing data from other sources like surveys, which we discuss in detail in Sect. 4.3.

In case of MMOs and other virtual environments, data can be divided into two main types: user characteristic data and user activity data. The user characteristic data consist of demographic characteristics, which are immutable in almost all cases like gender, age (which progresses at a constant rate) and semi-immutable characteristics constitute the characteristics that a user adopts for her virtual character, e.g., the character's gender, race, class, etc. The user activity data, on the other hand, consist of activities that a user performs in the gaming environment. While the user activity data can be said to capture all of the relevant that a person performs in a game without any filter, the same cannot be said about the user characteristic data since it is manually entered by the user. Thus it is possible for people to lie about their gender, age, location, etc. when they register to play a game. As an example, we consider the user data from EverQuest II described above where we observed that a subset of deviant players known as gold farmers always specified their location as either Alabama or Antarctica. The reason for choosing Alabama is likely because it is the first option to choose for a place of residence among the states in the USA. Such cases of obvious misdirect by the users have to be excluded from the analysis. Similar misdirects with respect to a person's age and gender can also happen and can be partially captured if additional sources of information for the same type of data are available.

When the game company is storing most activity data from its users, a number of data-related issues still have to be addressed. First, given the cost and storage and retrieval of large amounts of data the level of granularity at which the data are being saved can vary greatly. For example, should the telemetry information (location of a player character) in an MMO be saved at every microsecond or should it be saved whenever a player interacts with another player or a nonplayer character (NPC)? Notice that the former scheme is likely to generate massive amounts of telemetry data for games like WoW, which have millions of active players. Third, game logs and the corresponding database schemas are not designed with specific questions in mind. This can lead to poor performance with respect to data querying and retrieval.

4.2 Generalization

Issues related to generalization have been part and parcel of social sciences since their inception [13]. Traditionally, it has been extremely difficult to do experiments in the social science outside of the laboratory setting, which limits the size of the participant pool because of practical resource constraints. Generalization was thus dependent upon having enough observations over a number of populations which are varied enough [13]. There are also practical and ethical constraints with respect to manipulating human beings for the purpose of gathering data. In virtual worlds, however, since the cost of participation and making changes to the infrastructure is quite low as compared to their real world counterparts, it is possible to do things like AB testing on a massive scale. While virtual worlds offer a way to add at least partial intervention for data collection, they still leave open the question of replication across different virtual worlds. Consequently, the question of generalization *across* virtual worlds is still an unaddressed question as we shall argue below.

One of the principle issues with respect to studying MMOs is the lack of publically available datasets. Almost all the work that has been published in this area is either by scraping data from MMO websites or by researchers who get access to the datasets by working with the organizations that created the MMOs. As a consequence, access to the dataset is limited because of Non-Disclosure Agreements and confidentiality agreements. There are also legitimate liability and privacy concerns from the perspective of the gaming companies, which preclude sharing of the datasets as the Netflix data de-anonymization debacle demonstrates [20]. Thus as a result of these limitations, replication of results is limited and generalization of results is even more severely limited because researchers in general do not have access to multiple datasets. This can be especially problematic when researchers compare real world social interactions and those in the virtual world. The main epistemological risk is that generalizing results from one MMO or by comparing just one MMO with real world data.

A representative case of this issue is the work by Johnson et al. [16] on team formation in guilds by using WoW Data and the follow-up work by Ahmad et al. [3] on the replication of their results using EverQuest II data. Johnson et al. compared team data from guilds in WoW and street gangs in Los Angeles and proposed a model that can replicate the team size distribution in both these datasets. From their observations they concluded that a single mechanism is responsible for team formation in both offline and online settings. Ahmad et al. used the same model that was proposed by Johnson et al. and applied it to the guilds in EverQuest II. The results that were obtained by Ahmad et al. [3] were almost opposite to the results obtained by Johnson et al. The original model was predicated on the fact that the main driving force in team formation is the maximization of skillset but Ahmad et al. [3] were able to replicate these results by using a model that favors homophily, a mechanism explicitly ruled out by the original paper [16].

A number of lessons can be learned from this cautionary tale. One possibility is that it could be the case that the results from the WoW study are generalizable to the offline world but not from EverQuest II. The second possibility is that the convergence of results between WoW and the Los Angeles gangs' dataset was a fluke and the results are not generalizable from the study. The third possibility is that it is the EverQuest II dataset that corresponds well with the gang dataset. Yet another possibility is that there are yet undiscovered common generative mechanisms that describe team formation for all the three cases. Lastly, we note that the fifth possibility is that the gang in the offline world and the guilds in MMOs do not constitute a good mapping to draw any meaningful conclusions about the other. Regardless of which of these possibilities are correct, one observation that is common across all of these scenarios is that there are serious issues with respect to generalization unless data are employed to replicate the same test across multiple MMOs.

Even if we can do the mapping exercise correctly, it does not guarantee generalization. Thus claims regarding generalization based on one or two datasets should either come with the appropriate disclaimers or should be considered with extreme caution.

4.3 Surveys, Perception, and Truth

As described in the previous sections, even a perfect log database does not capture everything about a player's behavior. Information in the logs can be augmented with additional information by doing surveys of the players for whom the log data are already available. Additional sources of information can be added to game logs via social media websites like Facebook and Twitter, as well as online forums and communities, with the permission of the users. Information from these social networks can be used to augment in-world social network information between the users. Thus, veracity of certain types of information can be established to a greater degree if the same information is coming from multiple sources. For example, it could be the case a person is gender bending not just their virtual character but also reporting their real gender incorrectly in other environments. Thus in the EverQuest II data we also discovered instances where the real world gender reported by people on surveys is different from the real world gender that they reported when they signed up to play the game. Without an additional third source of information there is no way to disambiguate the gender of these players and thus these have to be left out of the analysis.

Out of these sources surveys can be a rich source of data to augment log data but survey data comes with its own set of unique problems. A number of studies spanning decades have shown that people do not always tell the truth on surveys [5, 23], either consciously or unconsciously, which calls the veracity of survey data partially into question. To illustrate this point, let us consider the example of amount of time a player spends playing games versus the amount of time that a player reports that he or she spends playing a game. In a study conducted by Williams et al. [24] it was discovered that both men and women underreport the time that they spend playing video games. Women actually underreport more as compared to men. The self-report information on the amount of time spent playing the game was determined from survey data, which was in turn linked to the game log data to infer the discrepancy between the reported hours and the actual number of hours spent. These results can be interpreted in a number of ways: it may be the case that people in general do not have a good handle on how much time they spend playing games. Or, people like to underreport because of some social stigma associated with spending too much time playing video games [25]. In both the cases, however, the result is the same—the time discrepancy remains. Thus any report which uses complementary survey data and limited options to add corrective measures should come with the appropriate disclaimers.

4.4 Mismapping from Virtual Worlds to the Real World

The Mapping Principle can only work if one knows what one is mapping to in the virtual world. In many cases, the virtual and the real will have many similarities and even share the same terminologies but it would be unwise to assume the two are equal. More often than not, one runs into examples where finding maps from one domain to the other but the conclusions from such studies are lacking because the mapping is not done properly. We refer to such cases as Mismapping. A common scenario where this happens is what the researchers do not invest enough time and effort in gaining expertise in the online domain (the virtual world or the MMO) that they are studying. Examples of such cases include misidentifying attributes of player characters [24] or misunderstanding of game mechanics [16]. The net effect of Mismapping is the tendency to make conclusions about the phenomenon.

Mismapping can, however, be easily avoided by investing some time and effort in actually playing the game in case of MMOs or spending some time getting familiar with the environment in the case of other virtual worlds. Usually spending 20 h in an environment seems to be a good guiding number for acquiring the minimal levels of expertise in such content-rich environments [26]. We would go as far as to recommend that conferences and journals should have a mandatory policy for requiring at least one author for any paper to have spent at least about a minimum of prespecified time in the virtual world that they are studying in order to avoid Mismapping. To illustrate this issue we again turn to the generalized team formation model of Johnson et al. [16]. In the paper, they assume that the guilds in WoW are analogous to ethnic groups in the street gangs in Los Angeles. This assumption is unwarranted because ethnicity based groups do not bear much resemblance to guilds in MMOs beyond a superficial level. Even if one were to make the argument admissible the organization of guilds in MMOs can range from something akin to playgroups to military style organizations [24] and would thus preclude any meaningful mapping between the two.

5 Case Studies: Virtual to Real World Mappings

There are a number of phenomena that are observed in the virtual world, which have their counterparts in offline worlds so that one can clearly observe similar types of behaviors in both the settings. The Mapping Principle [26] can be used to determine in which cases this mapping can be applied to these phenomena. In this section, we examine a few case studies where the Mapping Principle has been applied. We also investigate where mapping from the virtual to the real succeeds, and where it fails and the reasons for the failure.

5.1 Case Study: Virtual Economy

Economies of MGOs were one of the first phenomena to be observed in detail when such data started to become available. Both online and offline economies are characterized by finite resources and availability of resources which are driven by supply and demand. Certain games like EVE Online have actually hired professional economist to design the economy systems within [14]. A number of pioneering studies of the virtual worlds was done by Edward Castronova on the economies of virtual worlds [8–10]. It was observed that the virtual world economies exhibit many patterns that are observed in their real world counterparts. An added advantage of using virtual worlds as test beds for studying economic behavior is that they offer a way to simultaneously study both the microeconomic as well as macroeconomic behaviors of people. In the offline world it is not possible to do so because of privacy concerns and also because of the fact that data related to economic activities are spread over a vast number of sources. Virtual worlds also offer us a way to do AB testing in a manner that is not possible in the offline world. Even different economic systems with variants can be tried out to determine how people would act under different economic affordances and constraints.

The economies of the virtual worlds have many characteristics that one associates with a functioning economy: buying and selling of goods, creation and destruction of goods, banking, a parallel shadow economy in the form of gold farmers [1, 17], etc. Castronova et al. [10] studied the aggregate economic behaviors of players in EQ2 and compared the behaviors with real world economies and what theories of macroeconomic behaviors say about how humans should behave economically. The study reveal that aggregate economic measures like the GDP, price level, money supply, and inflation behaved in exactly the same way that their counter parts behave in the offline world, e.g., when the prices go up then demand decreases, when the population decreases which causes a decrease in demand then the prices go down. A natural experiment also occurred in their dataset where a new server was introduced in the game and the aggregate economic behavior of the new server quickly mimicked the aggregate behavior of the already existing servers.

Given the positive results the researchers suggested additional possibilities for future which can even have real world implications: researchers have traditionally used simulations to study the effect of changes in economic systems but virtual worlds can offer an even more realistic way in which people's reaction to an economic policy can be gauged. So before that change is introduced in the real world, policy-makers can try out such changes in the virtual world. The risks associated with implementing a policy in the virtual world with negative consequences would be far less as compared to implementing it in the offline world. Another intriguing part of

Fig. 1 The corrupted blood incident in WoW

the virtual economy is the clandestine black market economy that exists just like its offline counterpart [1, 17]. The main difference, however, is that in the offline world it is next to impossible to collect data about black market activities but such constraints are less severe in the virtual worlds [1]. Previous research has even shown that the behaviors of clandestine actors in the virtual worlds are very similar to their counterparts in the offline worlds [17].

5.2 *Case Study: Epidemiology*

The Corrupted Blood incident is a famous "global" event in WoW where a large number of players were affected by an unplanned in-game event [6] and seemed to offer potentially interesting insights to researchers of epidemiology. On September 13, 2005, Blizzard, the creators of WoW (Fig. 1), introduced a new dungeon instance into the game with a new boss associated with the dungeon. The new boss had a spell (Corrupted Blood) that acted like an infectious disease. Once a player was infected they could transmit the disease to other players and even NPCs if they were close enough. The spell damaged the "health" of the other characters over time. Blizzard had created the spell to be confined to the dungeon instance where it was introduced. But the unintended consequence of the spell is that it started to spread like a plague and eventually infected a large percentage of players in a number of servers [6]. A number of researchers in the field of epidemiology noted the similarities between how people reacted to Corrupted Blood in WoW and how people react to epidemics in the real world. Thus, Balicer [6] noted in one of the first papers on the subject, "a platform for studying the dissemination of infectious diseases, and as a testing ground for novel interventions to control emerging communicable diseases."

Beyond the obvious similarities between real world epidemics and the contagious nature of Corrupted Blood, researchers noted a number of other similarities: the Corrupted Blood originated in a remote, uninhabited region in WoW. It was carried by travelers to larger regions (Fig. 1) and also by players who were actively fleeing the main centers of the plague, the hosts for the plague were both human and animal [6]. In these respects some regarded that it had similarities to the Avian flu virus [6]. The similarities, however, end here and we should examine the efficacy of using virtual worlds to study these phenomena in the historical context. Earlier work in the field of Social Simulation focused on simulating influence and the spread of outbreaks but in a purely simulation-based framework where human agents are substituted for artificial agents [26]. Thus the main limitation of this approach was that certain assumptions that were made about human behavior like rationality or even bounded rationality were not borne out by psychological studies of human behavior [28]. The excitement of the research community sprung from the fact that these virtual environments represented spaces where one did not have to rely on virtual agents anymore and collect real data about human reactions to contagious diseases without any harm being done to humans.

On the other hand, the lack of liability on the part of the human beings is *also* a problem with respect to correct modeling of the phenomenon. The maximum penalty in WoW for contracting plagues is the death of the player's character, which can be regenerated. The player would at most lose some of the virtual resources acquired, but in the real world, a person who loses her life has no hope of ever being resurrected. People are likely to behave very differently when their lives are at stake. In conclusion, while there are a number of similarities between the two environments and the Mapping Principle seems to work well at first glance but the most important aspects of the two environments do not map and thus preclude us from making many useful conclusions about the real world in this case.

5.3 *Case Study: Deviant Clandestine Behaviors*

The phenomenon of deviancy is socially constructed. Thus, what may be considered deviancy in one context and culture may not be considered deviancy in another [1]. While deviant behaviors may or may not be clandestine, an important subset of such behavior is in fact clandestine in nature and these have mainly evaded analysis because of lack of data. It should be noted that clandestine activities may or may not be illicit since what constitutes illicit is also socially constructed. The common factor in the study of clandestine activities is that there is some effort by the parties involved to hide their activities. The Webster's dictionary defines clandestine activities as those activities which are "kept secret or done secretively, esp. because illicit." In the larger scheme of things it is extremely difficult to study deviant clandestine behaviors because of not just privacy implications but also the extreme difficulty to obtain data about criminal outfits. Here the Mapping Principle may be of benefit for, i.e., while the severity of deviant behaviors in the virtual worlds may be less, the constraints and affordances under which people operate in the two environments. With this observation in mind we consider various operationalizations of deviancy and clandestine behaviors in MMOs and what can be learned from them.

We revisit the WoW guilds example and the street gangs of Los Angeles [16]. The former represents a case of virtual organizations and the latter represents a possibly criminal organization, which has at least some elements of being clandestine in nature. In this case as well one cannot map the two environments because guilds in WoW are not clandestine in nature and street gangs are also semi-public in nature. To be fair, the authors of the paper do not make any claims regarding the deviant or the clandestine nature of the datasets. Thus, the affordances and the constraints of a clandestine environment do not map to the virtual world setting in the case of guilds. We now consider the example of gold farmers in MMOs; these are players who are involved in activities that are considered "illegal" by game administrators as they disrupt the economic balance of the game as well as challenge the assumption of the game as a meritocracy [1, 17]. Game administrators seek to actively ban gold farmers, and as a result the gold farmers change their behaviors to avoid detection [1]. This in turn forces the game administrators to update their strategies to catch gold farmers and then the cycle continues.

The gold farmer detection avoidance detection behavior is not restricted to simple behaviors but manifests in more complex ways, e.g., the gold farmers would extend their supply chains in order to put multiple layers of obfuscation between them and the game admins [1]. Gold farmers also avoid interacting with one another in modalities—for example, trust exists between pairs of gold farmers, which may indicate the presence of a strong relationship, but they interact with one another via proxies [17]. It has been noted that the avoidance detection techniques employed by the gold farmers [1], the thinning out of social networks as a tactic [17], etc., are very similar to how drug dealers operate. Thus a study by Keegan et al. [17] observed that there are a number of similarities between the social networks of gold farmers and the social networks of drug dealers, a dataset collected by the Calgary police department [17]. Keegan et al. [17] argue that the conditions under which the gold farmers operate are similar to the conditions under which drug dealers have to operate. In other words, in both cases the commodity being sold is limited in quantity, there is an active campaign by the authorities to clamp down on its distribution, there is a long supply line between the source and the distribution of the goods, important actors in the network placing themselves in noncentral positions where they are difficult to detect, etc.

The comparison by Keegan et al. [17] between the social networks of gold farmers and drug dealers revealed that these networks are much more similar to each other as compared to other social networks. At the surface, the two networks seem quite divergent but the presence of similar constraints and affordances reveal the presence of common generative mechanisms. This opens the possibility to study clandestine networks and behaviors in sufficient detail which is not possible in the offline world. We may be able to learn something about the offline counterparts of the virtual world deviant actors given a sufficiently careful study. A word of caution is however in order, it may be the case that the mapping in between the two environments works in the case of Keegan et al. [17] but the results may or may not be generalizable. Additionally, it is premature to base any policies based on the gold farmer and drug dealer comparison but one can still learn insights regarding their respective social networks given that the virtual world data are much rich in content [26].

5.4 Case Study: Mentoring

Mentoring is a phenomenon where a more experienced person in a given field serves as an adviser or a trainer to a less trained person [2]. There is an extensive amount of literature on mentoring in organizational theory [2] and small group research [2]. However, studies focusing on mentoring networks have been very limited and to date less than half a dozen such studies have been published [2, 4]. This is mainly owing to the fact that historically it has been extremely difficult in collecting mentoring data in a network setting. Prior work in this area can be divided into mentoring networks in the field of psychology and mentoring networks in MMOs. Temporal data for the psychology network is not available so that a meaningful comparison cannot be made between the two networks. Ahmad et al. [4] and Huffaker et al. [2] have done a series of studies on mentoring in MMOs. One of the things that they discovered was that the mentoring networks were different from most other social networks [4]. This of course raises questions regarding generalizability of not only social networks in MMOs but also domain-specific social networks. There are a number of motivations that people have with respect to mentoring others. The literature notes that people mentor because of instrumental reasons, friendship, organizational obligations, or a sense of paying it forward [2].

Starting with a set of features that characterized mentoring activities in the MMO EverQuest II, Huffaker et al. [4] observed that the most optimal clusters that they obtained corresponded to the four categories that are described in the literature. This observation also established at least some limited form of mapping between the virtual world and the offline world with respect to mentoring. The virtual world data, however, is also complemented with network information between the mentors and their apprentices. Huffaker et al. [4] observed that the various clusters of mentors also have different structural characteristics: instrumental mentors have a much higher value for closeness centrality as compared to other types of mentors, even though the instrumental mentors have a low value for mentoring. The explanation here is that the instrumental players are more focused on their own achievement but since they do not confine themselves to a small circle of friends their network spread out more. Also players who are focused on their virtual organizations or guilds have much higher clustering coefficients as compared to other types of mentors. Again this is expected, because mentors with this form of motivation naturally have greater engagement with other people. While most of these observations agree with what the researchers in the organizational studies' literature have suspected, the main lesson here is that since virtual worlds can offer us ways to learn about the real world and even test hypothesis which may not be possible in the offline world.

6 Predictive Modeling from Virtual World(s) to Real World(s)

In Sect. 1.3 we discussed the issues inherent in comparing the data-driven versus theory-driven paradigm. We note that the same debate has important implications not just for hypothesis formation and testing but also for predictive modeling. Consider a prediction task where the goal is to build models which also have explanatory power in terms of the features or the variables used. As an expositionary example, consider a behavioral prediction problem. There are two main ways in which such models can be constructed. We first consider the **theory-driven** paradigm: one can start with existing social science theories of human behavior and select or construct variables based on what the theories say about human behavior in that context. This can of course be augmented with new features which may appear in a new domain but which nonetheless fall with the purview of existing theory. Existing theories are also used to form hypothesis regarding the domain, these in turn are tested given the data. This is illustrated in Fig. 2 which also shows the alternative paradigm, the **data-driven** paradigm. In the data-driven approach one starts without any preconceived notions of how theory should drive model building. Models are built by features or variables which can be selected based on criteria like Information Gain, PCA, or other similar criteria. The process of model building itself is iterative in nature which may include things like parameter tuning, oversampling depending upon the distribution of classes, label propagation, etc.

It is important to note here that the two approaches are not exclusive in practice. There is always some element of data driving the direction of model building even in the case of theory building. And data-driven approaches may select different variables depending upon the domain where it is being applied even if feature sets and models themselves are not dependent upon the theory in the initial stages. However, treating their approaches as separate makes sense from a practical perspective. Consider the scenario given in Fig. 2 where the two approaches are being used to solve the same problem: four outcomes are possible. In the first scenario, the results from both the theory-driven and the data-driven approach agree, which implies that the data-driven results augment what the theory states. Thus, additional iterations of the model building process are not required. In the second scenario, one gets good results from the theory-driven approach but not from the data-driven approach. This would be the

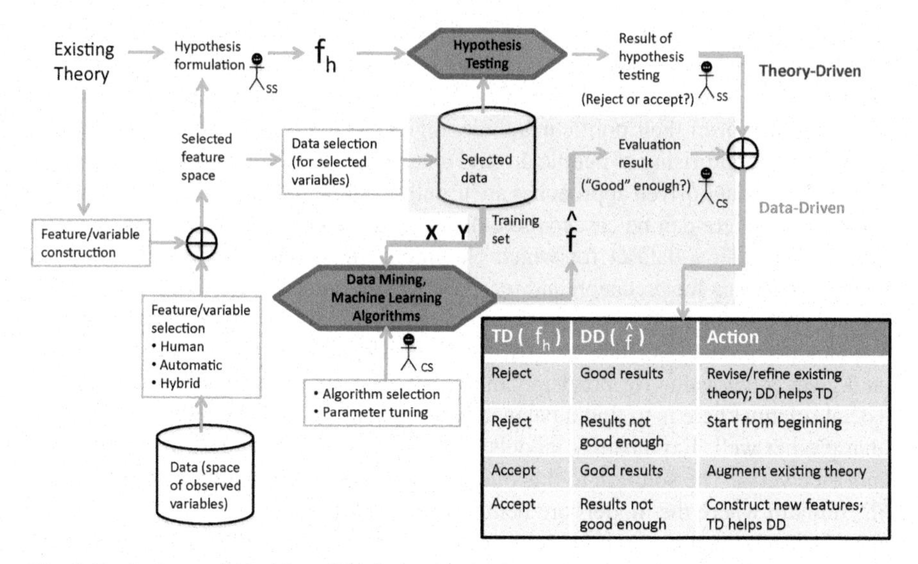

Fig. 2 Predictive models with social science theories

case where the data-driven models need to be updated based on insights from the theory. This may involve construction of new features from theory to augment model building. In the third scenario, one gets good results from the data-driven approach but not from the theory-driven approach. This is interesting because it implies that existing theory actually may have missed something and needs to be updated. This may involve updating existing theory with new explanatory variables, although the process may involve complications as we shall describe below. The last scenario is the trivial case where both the theory-driven and the data-driven approaches give weak results and the whole process of model building has to be revised from the beginning.

The particular approaches of actual model building for predicting real world from virtual world can vary depending upon the actual characteristics. The space of real world characteristics can be divided into ascribed characteristics and acquired characteristics [4]. Ascribed characteristics are the characteristics that are immutable in almost all the cases, such as gender, age (changes at a constant rate), personality-type (which changes rarely) [4], aptitude, intelligence quotient, emotional quotient, etc. Notice that these in turn can be divided into two subtypes: demographic characteristics and psychometric characteristics. The second class of characteristics is called acquired characteristics; these include education, location, vocation, etc. These are the characteristics of a person which can change over time thus a person who has a high-school diploma can get a bachelor's degree and have a different education status. For each of these variable types the choice of feature depends upon the domain as well as the variable that needs to be predicted. Consider the case of real world gender. Prior work has shown that while gender-bending is more common amongst men in virtual worlds [11], the gender of a person's primary character, i.e., the character with which they spent the most time playing is a good indicator of their real world gender. Similarly the choice of a character's virtual race and class seems to be a good indicator of their political and ideological leanings [26]. These observations should not be surprising as people do tend to take a part of them in the virtual worlds.

While the data-driven approaches are usually complementary to the theory-driven approaches, there can be cases where data-driven models may lack in explanatory power. Consider a dataset for which classifiers, like certain SVM [15, 22], may have high values for precision and recall but which are essentially black boxes that cannot shed much light onto why the model works. Now let us suppose that a more explanatory model like JRIP or Decision Tree [15] do not work well with a dataset and gives a low value for precision and recall. The dilemma for the computational social scientist here is to find a balance between explanatory models versus models that predict well. Explanatory models sometimes come at the cost of performance and vice versa. The solution to this dilemma then depends upon the application and the domain where the models are being applied. Thus if the goal is prediction then the black box nature of the model would be irrelevant but if the goal is to explain why the model works and how the various parts of the model fit together then one is left with no choice but to use the more explanation rich model at the expense of performance. While this tradeoff is not always needed, but it happens often enough that it is of note.

7 Conclusions

Virtual worlds have their origins in multi-user-dungeons in the late 1980s, which allowed multiple players to share the same gaming environment [1]. From these early environments they have evolved to become environments where millions of people can share these online persistent worlds. There are more than 400 million people who play MMO games in one form or another. This number itself represents 5.7 % of the population of the world and this number is likely to grow in the future. This implies the opportunities for further research in this area also abound. One can also think of the virtual world environments as natural experiments given the practical and logistical difficulties in collecting data about people in large assemblages. Given that many people who partake in virtual worlds do in fact spend significant time and effort in these virtual worlds, it should not be surprising that people's real world characteristics affect their gameplay and behaviors online. The Mapping Principle establishes mapping between virtual environments and their real world counterparts given that certain conditions are met. There are several examples where mapping may have been done based on insufficient similarities and thus prematurely mapped, which we caution as an example of how not to do computational social science. Another important point to keep in mind with respect to the Mapping Principle is that even if one can do the mapping exercise correctly it does not guarantee generalization. Results may not generalize to other MMOs let alone to the offline world.

The process of model building for predicting real world characteristics is iterative in nature and one can adopt data-driven as well as theory-driven approaches. These paradigms are complementary in nature and describe two different ways to conduct computational social science. One possible issue that can be problematic is the issue of tradeoff between explanatory powers of models versus their performance. In such scenarios, the decision regarding which models to use depends upon the demands of the application. The choice of variables that are used usually depends upon the domain and the context.

Acknowledgments Special thanks to Mushtaq Ahmad Mirza and Khalida Parveen for being there.

References

1. Ahmad, M.A., Keegan, B., Srivastava, J., Williams, D., Contractor, N.: Mining for gold farmers: Automatic detection of deviant players in MMOGs. In: Computational Science and Engineering, 2009. CSE'09. International Conference on (vol. 4, pp. 340–345). IEEE (2009, August)
2. Ahmad, M.A., Huffaker, D., Wang, J., Treem, J., Kumar, D., Poole, M.S., Srivastava, J.: The many faces of mentoring in an MMORPG. In: Social Computing (SocialCom), 2010 IEEE Second International Conference on (pp. 270–275) IEEE (2010, August)
3. Ahmad, M.A., Borbora, Z., Shen, C., Srivastava, J., Williams, D.: Guild play in MMOGs: rethinking common group dynamics models. In: Datta, A., Shulman, S. W., Zheng, B., Lin, S.-De, Sun, A., Lim, Ee-P. (eds.) Social Informatics, pp. 145–152. Springer Berlin Heidelberg (2011)
4. Ahmad, M.A., Ahmed, I., Srivastava, J., Poole, M.S.: Trust me, i'm an expert: Trust, homophily and expertise in MMOS. In: Privacy, security, risk and trust (passat), 2011 IEEE third international conference on and 2011 IEEE third international conference on social computing (socialcom) (pp. 882–887). IEEE ($62011, October)
5. Babbie, E.R.: Survey research methods. Wadsworth, Belmont (1990)
6. Balicer, R. D.: Modeling infectious diseases dissemination through online role-playing games. Epidemiology **18**(2), 260–261 (2007)
7. Borbora, Z., Srivastava, J., Hsu, K.W., Williams, D.: Churn prediction in MMORPGS using player motivation theories and an ensemble approach. In: Privacy, security, risk and trust (passat), 2011 IEEE third international conference on and 2011 IEEE third international conference on social computing (socialcom) (pp. 157–164). IEEE (2011, October)
8. Castronova, E.: Synthetic worlds: The business and culture of online games. University of Chicago Press, Chicago (2005)
9. Castronova, E.: On the research value of large games natural experiments in Norrath and Camelot. Games and Cult. **1**(2), 163–186 (2006)
10. Castronova, E., Williams, D., Shen, C., Ratan, R., Xiong, L., Huang, Y., Keegan, B.: As real as real? Macroeconomic behavior in a large-scale virtual world. New Media Soc. **11**(5), 685–707 (2009)
11. Davis, D.Z.: Gendered performance in virtual environments. Media Disparity: A Gender Battleground (2013): 133.
12. Einstein, A.: Out of my later years. Citadel Press, Secaucus (1956)
13. Fiske, D.W., Shweder R.A. (eds). Metatheory in social science: Pluralisms and subjectivities. University of Chicago Press, Chicago (1986)
14. Games, C.C.P.: Eve online. World Wide Web. http://www.eve-online.com (2003). Accessed 5 May 2014

15. Hall, M., Frank, E., Holmes, G., Pfahringer, B., Reutemann, P., Witten, I.H.: The WEKA data mining software: an update. ACM SIGKDD Explor. Newsl. **11**(1), 10–18 (2009)
16. Johnson, N.F., Xu, C., Zhao, Z., Ducheneaut, N., Yee, N., Tita, G., Hui, P.M.: Human group formation in online guilds and offline gangs driven by a common team dynamic. Phys. Rev. E. **79**(6), 066117 (2009)
17. Keegan, B., Ahmad, M.A., Williams, D., Srivastava, J., Contractor, N.: Dark gold: Statistical properties of clandestine networks in massively multiplayer online games. In: Social Computing (SocialCom), 2010 IEEE Second International Conference on (pp. 201–208). IEEE (2010, August)
18. Kandel, D.B.: Homophily, selection, and socialization in adolescent friendships. Am. J. Sociol. 427–436 (1978).
19. Lazer, D., Pentland, A.S., Adamic, L., Aral, S., Barabasi, A.L., Brewer, D., Christakis, N., Contractor, N., Fowler, J., Gutmann, M., Jebara, T., King, G., Macy, M., Roy, D. Van Alstyne, M.: Life in the network: The coming age of computational social science. Science (New York) **323**(5915), 721 (2009)
20. Narayanan, A., Shmatikov, V.: How to break anonymity of the Netflix prize data set. The University of Texas at Austin, Austin (2007)
21. Rogers, E.M., Bhowmik D.K.: Homophily-heterophily: Relational concepts for communication research.: Public. Opin. Quart. **34**(4), 523–538 (1970)
22. Vapnik, V.: The Nature of Statistical Learning. Data Min. Knowl. Discov. **6**, 1–47
23. Warner, Stanley L.: Randomized response: A survey technique for eliminating evasive answer bias. J. Am. Stat. Assoc. **60**(309), 63–69 (1965)
24. Williams, D., Ducheneaut, N., Xiong, L., Zhang, Y., Yee, N., Nickell, E.: From tree house to barracks the social life of guilds in world of Warcraft. Games Cult. **1**(4), 338–361 (2006)
25. Williams, D., Yee N., Caplan S.: Who plays, how much, and why? A behavioral player census of a virtual world. J. Comp. Mediat. Commun. **13**(4) 993–1018 (2008)
26. Williams, D.: The mapping principle, and a research framework for virtual worlds. Commun. Theory **20**(4), 451–470 (2010)
27. Yee, N., Bailenson, J.: The proteus effect: The effect of transformed self-representation on behavior. Hum. Commun. Res. **33**, 271–290 (2007)
28. Zastrow, C., Kirst-Ashman K.K.: Understanding human behavior and the social environment. Cengage Learning, Belmont (2007)

The Use of Social Science Methods to Predict Player Characteristics from Avatar Observations

Carl Symborski, Gary M. Jackson, Meg Barton, Geoffrey Cranmer, Byron Raines and Mary Magee Quinn

Abstract The purpose of this study was to investigate the extent to which real world characteristics of massively multiplayer online role-playing game (MMORPG) players can be predicted based on the characteristics and behavior of their avatars. Ground truth on participants' real world characteristics was obtained through the administration of validated measures of personality and authoritarian ideology, as well as a demographics form. A team of trained assessors used quantitative assessment instruments to evaluate avatar characteristics, behavior, and personality from a recorded session of the participant's typical gameplay. The statistical technique of discriminant analysis was then applied to create predictive models for players' real world characteristics such as gender, approximate age, and education level, using the variables generated through observational assessment of the avatar.

This work was supported by the Air Force Research Laboratory (AFRL). The US Government is authorized to reproduce and distribute reprints for Governmental purposes notwithstanding any copyright annotation thereon. Disclaimer: The views and conclusions contained herein are those of the authors and should not be interpreted as necessarily representing the official policies or endorsements, either expressed or implied, of AFRL or the US Government.

C. Symborski (✉) · G. M. Jackson · M. Barton · G. Cranmer · B. Raines · M. M. Quinn
Leidos, 4001 N, Fairfax Dr., Arlington, VA 22203, USA
e-mail: carl.w.symborski@saic.com

G. M. Jackson
e-mail: gary.m.jackson@leidos.com

M. Barton
e-mail: marguerite.r.barton@leidos.com

G. Cranmer
e-mail: geoffrey.cranmer@leidos.com

B. Raines
e-mail: byron.a.raines@leidos.com

M. M. Quinn
e-mail: mary.m.quinn@leidos.com

M. A. Ahmad et al. (eds.), *Predicting Real World Behaviors from Virtual World Data*, Springer Proceedings in Complexity, DOI 10.1007/978-3-319-07142-8_2,

1 Introduction

The prediction of an individual's real world (RW) characteristics, purely based on observations of his or her avatar's characteristics and behaviors in a massively multiplayer online role-playing game (MMORPG), is a challenging yet fascinating proposition. The complex relationship between players and their avatars is not yet fully understood. While some players prefer to approximate their RW personas and appearances in virtual worlds (VWs), or even present more attractive, idealized versions of themselves [1], [2], others might opt to present a role-playing identity in-world [3]. This makes the determination of relationships between the avatar and his/her operator a unique challenge. A player may be presenting physical or behavioral characteristics consistent with his/her RW persona, or his/her avatar appearance and behaviors may display characteristics entirely different from that persona.

With this in mind, a study was designed to investigate whether it is possible to predict RW characteristics of MMORPG players from the characteristics and behavior of their avatars. Other studies, such as [4], have sought to accomplish the same end by parsing the massive amount of player data logs stored in MMORPG databases and tying the results back to measured RW player characteristics. This study sought to use more traditional modes of social science research and statistical techniques to develop predictive models, with avatar characteristics and behavior as the predictor variables and RW player characteristics as the criterion, or dependent, variables. This facilitated the creation of relatively simple algebraic equations that could be used by a trained observer to make predictions about an individual after a period of observing his/her avatar.

Using quantitative predictor/independent variables (IVs) generated through the observational collection of avatar data and criterion/dependent variables (DVs) collected through the administration of several ground truth instruments filled out by participants, the statistical technique of discriminant analysis (DA) was then employed to produce predictive models of RW characteristics using VW observations. In this study, the quantitative variables gathered were used to generate statistical models for the prediction of RW gender, age, education level, extraversion level, and submissive ideology.

2 Method

2.1 Participants

Participants were recruited using online advertisements, flyers, and online game forums. All participants were required to be at least 18 years of age and to have a minimum of 50 h of experience with either *Guild Wars®* [5] or *Aion®* [6], two popular MMORPGs. A total of 114 participants completed the study. Recruitment occurred primarily within the Washington DC region, USA.

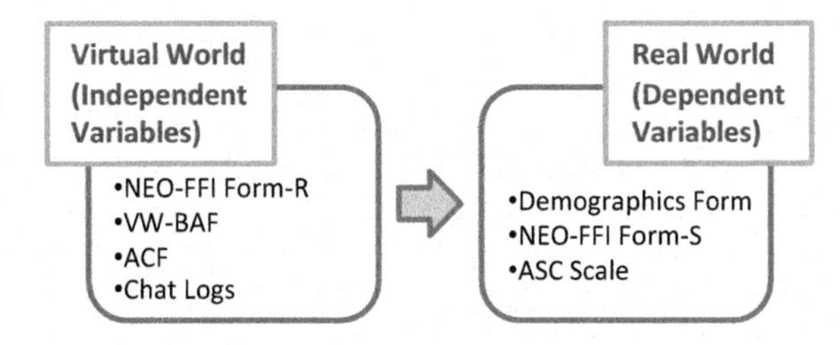

Fig. 1 Sources of real world dependent variables (DVs) and virtual world independent variables (IVs)

More *Guild Wars*® (73, or 64 %) than *Aion*® players (41, or 36 %) and more males (71, or 62 %) than females (43, or 38 %) participated in the study. The mean participant age was 31.7 years' old, with a standard deviation of 9.5 years. A slight majority of participants had less than a 4-year degree (51, or 45 %), while 46 participants held a Bachelor's degree (40 %), and 17 participants (15 %) held a graduate degree.

2.2 *Instruments*

An important facet of the research design was the distinction between the DVs to be predicted, or the selected RW characteristics—gender, age, education level, extraversion level, and submissive ideology—and the IVs to be used as predictors for the selected DVs, or the characteristics and behavior of the players' avatars. Several instruments were used to capture these data (see Fig. 1).

2.2.1 Data Collection Instruments—RW Characteristics

The RW characteristics of the participants, or the DVs to be predicted, were recorded using three forms. The **Demographics Form** was used to collect basic demographic information on the participants, such as gender, age, and highest education level achieved. Each participant also completed the validated **Aggression-Submission-Conventionalism Scale (ASC scale)**, which has three subscales—submissive, aggressive, and conventional—as a measure of ideology [7].

Finally, each participant completed the **NEO Five-Factor Inventory (NEO-FFI),** a standardized and validated personality assessment instrument constructed from the five-factor model of personality [8]. The instrument measures the "Big Five" traits of neuroticism, extraversion, openness to experience, agreeableness, and conscientiousness. The NEO-FFI was selected for use because it has two forms: the NEO-FFI Form-S (NEO-S), a self-report form, and the NEO-FFI Form-R (NEO-R), an observer rating form; the two forms are identical, except that the NEO-S is written from a first person perspective and the NEO-R is written from a third person perspective [8]. This enabled the use of the NEO-S to gather ground truth data on the participants' RW personalities, and the use of the NEO-R to gather objective assessments of the avatar's personality, made by an independent rater (see below).

2.2.2 Data Processing Instruments—Avatar Characteristics and Behavior

Multiple mechanisms for collecting IV data on the characteristics and behaviors of the participants' avatars were employed. The NEO-R was used to gather information on avatar personality, the Virtual Worlds Behavior Analysis Form (VW-BAF) and Avatar Characteristics Form (ACF) were developed to collect data on avatar behaviors and characteristics, and chat logs were analyzed for linguistic data.

While the NEO-S was completed by the participants, the **NEO-R** (observer rating form) [8] was used to assess avatar personality. The NEO-R was completed by trained assessors observing the participant's avatar, facilitated by a "bridge" developed to resolve inconsistencies or conflicts of any NEO-R item that caused rating difficulty within the VW, considering that the original form had been developed to rate humans in the RW.

The **VW-BAF** was created to measure avatar activities and behavior in MMORPGs. The form has two pages. The first allows a rater to tally occurrences of behaviors on a minute-by-minute basis (e.g., number of emotes, number of times avatar dies, number of occasions avatar helps another avatar). The second page allows a rater to record presence or absence of major activity categories (e.g., quest activity, social activity) during the interval. The second page of the form also measures how much time an avatar spends in a group versus solo, and whether the mode was Player versus Player (PvP) or Player versus Environment (PvE).

The **ACF** was developed to capture the following general characteristics of avatars in the VW: gender, physical characteristics, class/role, armor/weapons, style of play (e.g., member of a guild), pets, and general play style (e.g., social player, strategic player).

A central feature of the instrument development was that all forms—the NEO Bridge, the VW-BAF, and the ACF—were designed to be generalizable across all MMORPGs. This is important because MMORPGS have a wide variety of different naming conventions for similar things (e.g., a healer class may be called a "monk" in one game and a "chanter" in another). Thus, the adoption of a standard vernacular was an essential component of the form development process. For example, regardless of game nomenclature, all character classes were assigned to one of the following four categories based on their primary characteristics: warrior, scout, priest, and mage.

Additionally, the research team transcribed participant **chat logs**. Care was taken only to record chat by the target participant/avatar. These chat logs were run through a parser, which extracted variables such as number of words and average word length.

2.3 Design and Procedure

The basic research design included three primary phases which will be described here: the initial laboratory session with the participant to establish ground truth, the gathering/processing of quantitative data, and the development of predictive statistical models using DA.

2.3.1 Laboratory/Home Session

Upon arriving at the laboratory, each participant viewed a standard introductory video describing the study and completed the informed consent process, if he/she desired to participate. The participant then completed the demographics form, the NEO-S, and the ASC scale. These collected data formed the basis for ground truth on the participant's RW characteristics.

At the participant's leisure after returning home, the participant recorded a 1-h gameplay session of the selected game using a screen capture program. Participants were requested to choose one avatar and to use only that avatar for the entire 1-h home session, and to play the game just as they normally would.

Once the home session was completed, the participant saved the recorded session on a USB flash drive and mailed it back to the laboratory. Participants were asked to complete the home session within 2 weeks of the laboratory session. Once the flash drive was received, participants were compensated for their contribution to the study.

2.3.2 Quantitative Data Processing

Upon receipt of the flash drive by the laboratory, trained assessors rated that session using the NEO-R, the VW-BAF, and the ACF. Separately, chat was transcribed. A minimum of 80 % inter-rater agreement was obtained for all rating forms.

For completion of the NEO-R, raters observed the entire hour of the recorded session and then provided avatar ratings using the NEO Bridge to help answer specific items, such as "He often feels tense and jittery."* For this item, the NEO Bridge specified that, if the avatar exhibited spastic movement, swiveled the camera frequently, ran in circles or rapidly manipulated the user interface, the item was coded as "agree." However, if the avatar typically appeared relaxed and calm, the item was coded as "disagree."

The VW-BAF rating occurred across three 10-min samples, which were standardized in terms of start times across the full 1-h session to ensure a representative sample of avatar behavior. Observations were recorded once a minute. As described previously, the VW-BAF has two pages: the first is a series of counts for how many times particular behaviors occurred each minute, and the second recorded whether or not certain behaviors occurred within each minute. Both kinds of variable were summed over each 10-min sample, and then averaged across the three 10-min samples for each participant before being used for analysis. Hence, each VW-BAF variable either corresponds to how many times a behavior occurred in an average minute, or to how many minutes out of 10 included that behavior on average.

At the same time that the VW-BAF ratings were recorded, the ACF was completed. The ACF items consist of presence or absence variables; for example, presence or absence of a costume or of a ranged damage-per-second combat role.

2.3.3 Statistical Technique: DA

DA was used to generate predictive models for participants' RW characteristics based on their avatars' characteristics and behavior. The purpose of DA is to predict group membership, based on RW DVs such as gender or extraversion, from a linear combination of VW IVs, such as avatar gender or type of armor worn. DA begins with a data set containing many cases (participants), where both the values of the IVs and the group membership (DVs) are known. The end result is an equation or set of equations that predict(s) group membership for new cases where only the values of the IVs are known [9].

Specifically, DA using *backward* stepwise reduction was conducted. This form of stepwise reduction begins with a given set of variables and reduces the set by eliminating the variables that are associated with the DV to a lesser degree than remaining variables [10]. In this way, only the variables that are most predictive of the DV remain as part of the predictive model when the analysis is complete. As an additional quality control, leaving-one-out cross-validation was used for all stepwise reduction DA runs. This form of validation accuracy is the process by which a model is trained on all cases but one and tested on the one case that was withheld. The process repeats until all cases have been withheld and tested blindly, eliminating any chance of predicting a case based on extracted knowledge of that case [10].

3 Results and Discussion

3.1 The Discriminant Function

As described above, DA generates a predictive model in the form of a linear combination of independent, or predictor, variables. There is also a constant term for each equation, which is used as the linear offset in the discriminant functions. The general form of the discriminant function is as follows:

$$DV\ Group\ 0 = (b_1 \times x_1) + (b_2 \times x_2) + \ldots + (b_\mathrm{n} \times x_\mathrm{n}) + c,$$

$$DV\ Group\ 1 = (b_1 \times x_1) + (b_2 \times x_2) + \ldots + (b_\mathrm{n} \times x_\mathrm{n}) + c,$$

where,

$b_n =$ the Fisher coefficient for that variable,
$x_n =$ the value of the independent variable, and
$c =$ the value of the constant.

Once the values of the variables are substituted in the above equations for the names of the variables, whichever of the two equations evaluates to the greater number will be the prediction for that participant. In other words, whichever of the two Fisher's discriminant functions produces a higher value "wins," and the participant will be predicted as a member of the corresponding category [11].

Table 1 Overall accuracy for predicting RW characteristics from VW observations

RW characteristic	Overall accuracy (%)	Precision (%)	Recall (%)
Gender	83	93	79
Age	70	80	56
Education Level	66	68	71
Extraversion Level	68	64	63
Submissive Ideology	65	66	66

RW real world, *VW* virtual world

3.2 Accuracy Metrics

Several accuracy metrics are reported for each of the predictive models. The first is the leaving-one-out cross-validated overall accuracy for each model. It is the best representation of the DA model because it captures information about how well the model classifies cases into both the target and non-target groups. Precision, representing the proportion of those cases predicted to be in the target group that were actually in that group, and recall, denoting the proportion of cases in the target group that were correctly predicted to be in the target group, are also presented to provide information about the accuracy of the models.

3.3 Overall Results

Table 1 presents the accuracy results of the developed models. Individual results are presented in the following sections. These sections explain each model along with a description of how the DV was defined, an overview of the accuracy of the model, and a brief discussion of the IVs relevant to the prediction of the target DV. Some of these discussions of IVs have more obvious support in the literature and appear to be more clear-cut than others. For the cases in which the relationship between the predictor IV and the DV being predicted was not clear, the subject matter experts of the research team developed potential explanations for the relationship based on their knowledge of MMORPGs.

3.4 Gender

3.4.1 Definition of the DV

For the DV of gender, males were assigned as one group (71 participants) and females were assigned as the other group (43 participants).

Table 2 Accuracy of gender model

Overall accuracy (%)	Precision (%)	Recall (%)
83	93	79

3.4.2 Accuracy of Gender Model

The gender model obtained 83 % overall accuracy (see Table 2). Precision was high at 93 % and recall was 79 %.

3.4.3 Discriminant Function for Gender

In determining which group a new case should be classified into, the values for each of the IVs were plugged into both of the following two equations. Whichever equation yielded the largest value represents the group that the new case was classified into. The following are the discriminant functions derived for gender:

$$\begin{aligned} Male = {} & (4.800 \times MaleAV) + (4.924 \times MajRoleSupport) \\ & + (2.429 \times HairAccNA) - 2.646, \end{aligned} \tag{1}$$

$$\begin{aligned} Female = {} & (0.578 \times MaleAV) + (0.695 \times MajRoleSupport) \\ & + (0.081 \times HairAccNA) - 0.721, \end{aligned} \tag{2}$$

Table 3 presents descriptions of the predictor, or independent, variables in the discriminant functions.

Through interpretation of the discriminant functions above, the models can be described loosely as follows:

If the avatar is male, and/or has a majority combat role of support, and/or has covered hair, then it is likely that the player's RW gender is male.

Otherwise, it is likely that the RW gender is female.

NOTE: These written descriptions are not to be substituted for the above equations in making predictions; they are only intended to aid the reader in understanding the statistical models.

3.4.4 Discussion of IVs Relevant to the Prediction of Gender

Avatar Gender (ACF)

It is well-known that avatar gender is closely related to RW gender [12]; therefore, it is not surprising that avatar gender surfaced as a predictor variable in this analysis. In our study worlds, the appearance of a male avatar strongly predicted a RW gender of male. Female avatars required the support of additional IVs to distinguish between female avatars operated by RW females and female avatars operated by RW males, because "gender-bending" is more common among male players (who play female avatars) than among female players [12].

Table 3 Description of IVs relevant to prediction of gender

Variable	Description of variable
MaleAV	Avatar is male
MajRoleSupport	Majority combat role is support (e.g., healing, buffing)
HairAccNA	Hair accessories cannot be observed, likely because hair is covered (e.g., by a helmet or costume)

IVs independent variables

Majority Role of Support (ACF)

This variable indicated whether or not an avatar's majority combat role was "support," a combat style characterized by healing, providing enhancement buffs, or otherwise supporting others in battle (50 % of the time or greater). It is commonly assumed that females gravitate toward healing roles in MMORPGs, since women are considered to be more nurturing and supportive by nature [13]. Our research suggests the opposite; the presence of this theme was a predictor of RW male gender. Because male avatar gender is such a strong predictor of RW male gender, the remaining variables in the discriminant function above primarily indicate the remaining RW males who were "gender-bending." Yee et al. [14] conducted a study that offers an explanation as to why the theme of healing others in combination with the presence of a female avatar might predict RW males: they discovered that "[male] players enact this stereotype [of women as healers] when gender-bending." Hence, a male player creating an avatar for a healing role might tend to choose a female.

Covered Hair (ACF)

The variable "Hair Accessories N/A" indicated if an avatar's hair accessories could not be observed because the avatar's hair was covered, due to a helmet or costume that occluded the hair (e.g., a hood). The model indicated that RW male players playing female characters in the sample were less likely to expose their hair and hair accessories than RW female players were. While there does not appear to be a ready explanation for this phenomenon in the literature, the research team postulated that RW male players operating female avatars might be less concerned with displaying their hair than RW female players, instead being more focused on gameplay activities, while RW female players may have more interest in ensuring that their avatars have a feminine appearance, complete with displayed flowing locks.

3.5 Age

3.5.1 Definition of the DV

The DV of age was divided into two groups: under the age of 30 and 30 and over. This is consistent with demographic research [15] and qualitative observations indicating

Table 4 Accuracy of age model

Overall accuracy (%)	Precision (%)	Recall (%)
70	80	56

that MMORPG players tend to be younger than social game players. Furthermore, the age group of 18–29 years is one that is frequently cited in research studies, particularly in voting-related and medical contexts, and thus seemed a logical breakdown to use. Using this division, 59 participants were age 30 years or over and 55 participants were under age 30.

3.5.2 Accuracy of Age Model

The age model achieved 70 % overall accuracy (see Table 4). Though precision was high at 80 %, recall was low at 56 %, suggesting that this model may need to be further refined.

3.5.3 Discriminant Function for Age

The following are the two equations derived for age:

$$30\ \mathit{or\ Older} = (1.206 \times \mathit{BAF35}) + (2.789 \times \mathit{UnconvHair}) + (2.245 \times \mathit{NEOR21}) - 2.814, \tag{3}$$

$$\mathit{Under}\ 30 = (0.397 \times \mathit{BAF35}) + (0.815 \times \mathit{UnconvHair}) + (2.905 \times \mathit{NEOR21}) - 3.008. \tag{4}$$

Table 5 presents descriptions of the variables used.

Simply stated, this model becomes:

If the avatar spends more time stationary, and/or has unconventionally colored hair, and/or does not appear tense and jittery, then it is likely that the player's age is 30 or over.

Otherwise, it is likely that the player's age is under 30.

3.5.4 Discussion of IVs Relevant to the Prediction of Age

Avatar does not move for the full 60 s (VW-BAF)

This item measured how many full minutes an avatar spent stationary over the observation period. In the study sample, an avatar that spent more time stationary was more likely to be older (age 30 or over). Younger people tend to be more active than older people, and perhaps more prone to bouts of fidgeting and nervous activity. In

Table 5 Description of IVs relevant to prediction of age

Variable	Description of Variable
BAF35	Avatar does not move for the full 60 s
UnconvHair	Avatar's hair is either an unnatural color (e.g., blue) or white
NEOR21	She often feels tense and jittery.*

IVs independent variables

the VW, this may translate into moving frequently, walking in circles, or jumping repeatedly, and an avatar that is frequently in motion is unlikely to be stationary for a full minute at a time. Conversely, an avatar operated by an older individual is more likely to spend time stationary.

Unconventional Hair (ACF)

This item indicated if an avatar's hair was either unnaturally colored (e.g., blue, green) or white. In the study sample, an avatar with unnaturally colored or white hair was more likely to be operated by a player age 30 or older. The selection of white hair for an avatar is consistent with previous research which shows that older players like to be able to make their avatars look older and, given limited customization choices, white hair is one of the few mechanisms that they have to do so [16]. The choice of unnaturally colored avatar hair (e.g., blue hair) by older players is not as easily explained, based on the available literature. The research team hypothesized that older players may be less concerned about conforming to social norms, or may be experimenting with representations that they would not try in the RW. While younger people may feel as though it would be acceptable to have unconventionally colored hair in the RW, older players may feel that the only acceptable place for them to experiment with unconventional hair color is in the VW.

"She often feels tense and jittery."* (NEO-R)

This personality assessment item was completed after an avatar was observed for the 1-h gameplay interval. If the avatar exhibited spastic movement, swiveled the camera frequently, ran in circles or rapidly manipulated the user interface, the item was coded as "agree," as directed by the NEO Bridge. Alternatively, if the avatar typically appeared relaxed and calm, the item was coded as "disagree." In the study sample, an avatar that was observed to be typically relaxed and calm (not tense and jittery) was more likely to be age 30 or over. This item is synergistic with the previous variable "Avatar does not move for the full 60 s," in that older players were more likely to be calm and stationary while younger players were more likely to be in motion and fidgeting.

Table 6 Accuracy of education level model

Overall accuracy (%)	Precision (%)	Recall (%)
66	68	71

3.6 Education Level

3.6.1 Definition of the DV

The DV of education level was divided into two groups: less than a Bachelor's degree (51 participants) and Bachelor's degree or greater (63 participants).

3.6.2 Accuracy of Education Level Model

The overall accuracy for the education level model was 66 % (see Table 6). Precision and recall were fairly balanced at 68 % and 71 %, respectively.

3.6.3 Discriminant Function for Education Level Model

The following are the two equations derived for education level:

$$\begin{aligned} \mathit{BA\ or\ Greater} &= (1.356 \times \mathit{RoleCtrlRanged}) \\ &\quad + (0.672 \times \mathit{BAF4}) - 0.808, \end{aligned} \tag{5}$$

$$\mathit{Less\ than\ BA} = (2.607 \times \mathit{RoleCtrlRanged}) + (1.855 \times \mathit{BAF4}) - 1.695. \tag{6}$$

Table 7 presents descriptions of the variables used.

Simply stated, this model becomes:

If the avatar does not have a combat role of ranged control and/or does not curse or insult others, then it is likely that the player holds a Bachelor's degree or greater.

Otherwise, it is likely that the player does not have a Bachelor's degree.

3.6.4 Discussion of IVs Relevant to the Prediction of Education Level

Role of Ranged Control (ACF)

This variable indicated whether or not the avatar was observed filling a combat role of "ranged control," which is a style of controlling opponents from a distance (e.g., crowd control). In the study sample, an avatar that did not exhibit a combat role of ranged control was more likely to have a Bachelor's degree or greater. In the opinion of the research team, the combat roles of ranged control and ranged damage-per-second tend to be the most high-powered and easiest to master, whereas combat

Table 7 Description of IVs relevant to prediction of education level

Variable	Description of variable
RoleCtrlRanged	Player engages in the combat role of ranged control (e.g., crowd control)
BAF4	Number of occasions of exhibiting curse words, insults, or "put downs" as directed at others

IVs independent variables

Table 8 Accuracy of extraversion level model

Overall accuracy (%)	Precision (%)	Recall (%)
68	64	63

roles such as healing or tanking are more difficult to master and are therefore more cerebral. That more educated players were less likely to perform ranged control suggests that perhaps they tended to gravitate toward the more cerebral healing or tanking roles instead.

Number of curse words or insults (VW-BAF)

This variable indicated the extent to which an avatar was observed using curse words or insulting others during the period of observation. In the study sample, an avatar that did not curse or insult others was more likely to be operated by a player with a Bachelor's degree or greater. This suggests that cursing or insulting others may be characteristic of less educated players.

3.7 *Extraversion Level*

3.7.1 Definition of the DV

The NEO scoring system groups individuals into very high, high, average, low, and very low categories of extraversion [8]. To split the sample into groups, those with high or very high extraversion were placed in the high extraversion group (51 participants), and those with very low, low, or average extraversion were placed in the low extraversion group (63 participants).

3.7.2 Accuracy of Extraversion Level Model

For the extraversion level model, 68 % overall accuracy was achieved (see Table 8). Precision was 64 % and recall was 63 %.

Table 9 Description of IVs relevant to prediction of extraversion level

Variable	Description of variable
NEOR7	He laughs easily.*
RoleDPSRanged	Player engages in the combat role of ranged DPS
NEOR14	Some people think he is selfish and egotistical.* [reverse-scored]

IVs independent variables

3.7.3 Discriminant Function for Extraversion Level

The following are the two equations derived for extraversion level:

$$\begin{aligned} \textit{High Extraversion Level} = {} & (1.234 \times \textit{NEOR7}) + (2.012 \times \textit{RoleDPSRanged}) \\ & + (3.281 \times \textit{NEOR14}) - 6.586. \end{aligned} \tag{7}$$

$$\begin{aligned} \textit{Low Extraversion Level} = {} & (0.757 \times \textit{NEOR7}) + (2.953 \times \textit{RoleDPSRanged}) \\ & + (3.747 \times \textit{NEOR14}) - 7.328. \end{aligned} \tag{8}$$

Table 9 presents descriptions of the variables used.

Simply stated, this model becomes:

If the avatar laughs easily, and/or does not have a combat role of ranged DPS, and/or appears selfish and egotistical, then it is likely that the individual has high extraversion.

Otherwise, it is likely that the individual has low extraversion.

3.7.4 Discussion of IVs Relevant to the Prediction of Extraversion Level

"He laughs easily."* (NEO-R)

This item was the rater's evaluation of whether or not the avatar laughed easily. A rater agreed with this item if the avatar laughed within the 1-h gameplay session (e.g., "lol," "haha," audible laughter), and disagreed with this item if the avatar did not laugh, as specified on the NEO Bridge Form. In the study sample, an avatar that laughed was more likely to be operated by a highly extraverted individual, which makes sense because this NEO-R item is part of the extraversion subscale. It is interesting to note that this facet of RW extraversion generalizes to the VW.

Role of Ranged DPS (ACF)

This item captured whether or not the avatar engaged in a role of ranged DPS, characterized by initiating damage attacks from a distance to the target (e.g., shooting the enemy with a bow from a distance away), during gameplay. In the study sample, not having a ranged DPS combat role helped to predict a high extraversion level in the RW. Previous research has indicated that tanking and melee DPS players tend to "lead the charge" into battle and otherwise direct the group on attack strategy [13]; therefore, it might make sense that a highly extraverted, leader-like individual would be drawn to tanking/melee DPS roles rather than ranged DPS roles.

"Some people think he is selfish and egotistical."* (NEO-R)

This item was the rater's evaluation of whether or not the avatar appeared selfish and egotistical during gameplay. A rater agreed with this item if the avatar disregarded requests or commands, prioritized his/her needs over those of others (e.g., looting during battle), or bragged about his/her accomplishments to other avatars, as specified on the NEO Bridge Form. A rater disagreed with this item if the avatar appeared selfless and humble (e.g., followed commands, helped others). In the study sample, selfish and egotistical behavior helped predict high extraversion in the RW. It may be that extraverted individuals, who frequently emerge as leaders in groups [17], may appear selfish and egotistical because they are more concerned with commanding others for their own purposes and getting their own way than with being selfless and humble.

3.8 *Submissive Ideology*

3.8.1 Definition of the DV

An item from the ASC scale was used to measure a submissive ideology: "ASC 2. Our leaders know what is best for us" [7]. This item was selected on the basis that it was the most representative item from the ASC authoritarian submissive subscale. The idea that DV groups could be formed using a specific item from an assessment scale was considered to be particularly interesting; by achieving accuracy in the prediction of submissive ideology, it is possible to demonstrate the prediction of an individual's response to a specific statement. Those participants who responded neutral, agree, or strongly agree to the item were considered submissive (59 participants); those who responded disagree or strongly disagree were considered not submissive (55 participants).

3.8.2 Accuracy of Submissive Ideology Model

For this model, 65 % overall accuracy was obtained (see Table 10). Precision and recall were equivalent at 66 %.

Table 10 Accuracy of submissive ideology model

Overall accuracy (%)	Precision (%)	Recall (%)
65	66	66

3.8.3 Discriminant Function for Submissive Ideology Model

The following are the two equations derived for submissive ideology:

$$\textit{Submissive Ideology} = (2.642 \times \textit{NEOR24}) + (-3.240 \times \textit{PvP}) - 3.850, \quad (9)$$

$$\textit{No Submissive Ideology} = (1.961 \times \textit{NEOR24}) + (-0.291 \times \textit{PvP}) - 2.470. \quad (10)$$

Table 11 presents descriptions of the variables used.

Simply stated, this model becomes:

If the avatar is not cynical and skeptical of others' intentions and/or does not engage in PvP, then it is likely that the player has a submissive ideology.

Otherwise, it is likely that the player does not have a submissive ideology.

3.8.4 Discussion of IVs Relevant to the Prediction of Submissive Ideology

"She tends to be cynical and skeptical of others' intentions."* (NEO-R)

This item was the rater's evaluation of whether or not the avatar was "cynical and skeptical of others' intentions." A rater agreed with this item if the avatar voiced doubt; complained about the game's programming or other players; or exerted influence over others' behavior, particularly by planting flags or issuing commands. A rater disagreed with this item if the avatar did not exert influence over others' behavior with commands and/or did not express complaints or doubts about others. Instead, the avatar might have followed others or complimented other avatars or the game. In the study sample, not being cynical and skeptical of others' intentions helped predict a submissive ideology. A submissive person would likely not be cynical and skeptical of others' intentions, particularly since the ASC scale item that was used to define this DV—"Our leaders know what is best for us" [7]—is clearly devoid of cynicism or skepticism.

Player engages in PvP (ACF)

This item indicated whether or not the avatar engaged in PvP mode at any time during the observation period. In the study sample, not participating in PvP helped to predict an RW submissive ideology. The research team theorized that a submissive person would not to be particularly interested in engaging in PvP combat, which requires a dominant and aggressive attitude.

Table 11 Description of IVs relevant to prediction of submissive ideology

Variable	Description of variable
NEOR24	She tends to be cynical and skeptical of others' intentions.* [reverse-scored]
PvP	Player engages in PvP

IVs independent variables

4 Conclusions

The aim of this study was to develop models to predict an individual's RW characteristics from his or her avatar behavior in the VW. To that end, the research team generated statistical models for five RW characteristics—gender, age, education level, extraversion level, and submissive ideology—from quantitative measures of avatar personality, behavior, and characteristics. The overall accuracy across the five models suggests that it is possible to develop predictive models for RW characteristics from observations of VW characteristics and behavior.

Beyond the predictive models themselves, this chapter presents the innovative use of quantitative instruments to gather predictor variables for use in the modeling process. To our knowledge, the application of the NEO-R, bridged for use as an assessment of avatar personality in the VW, is a new approach in the field of VW research. A second innovation was the development of a VW behavior assessment instrument (VW-BAF) that standardized ratings of avatar behaviors and allowed for reliable behavioral measures across all avatars. Additionally, the research team consistently recorded observable avatar characteristics via the ACF. These three quantitative measures proved invaluable in providing predictor variables for the statistical modeling process.

The research team was interested to note that there was a complete absence of chat variables from any of the predictive models, which spurred an exploratory, qualitative investigation of the chat data. Two conclusions were reached: first, in the sample, many participants did not chat at all, and those who did spoke very little, with most conversation content limited to current game activities. Second, qualitative analysis combined with demographic data allowed the research team to determine that the use of all lowercase letters and absence of punctuation was a universal behavior pattern that transcended RW demographic categories such as age and education. In our sample, these findings may explain why the predictive models were devoid of chat variables that could discriminate between groups.

There were several limitations in this study. First, the small sample size ($N = 114$) may be an impediment to the statistical power of the model development. The requirement to bring each participant into the laboratory, while it enhanced certainty in the validity of the ground truth data, limited recruitment—both geographically and culturally, in the sense that only MMORPG players willing to come in for a laboratory session participated. Second, the observation of only 1 h of recorded gameplay was less than ideal. Due to time and staffing constraints associated with the rating process, it would have been very difficult within the scope of this study to process more participant data.

Future research could benefit from the use of quantitative assessment instruments such as the NEO Bridge, the VW-BAF, and the ACF. Furthermore, future research should seek to validate the predictive models developed in this study with larger sample sizes and by rating the participants over multiple hour-long observation sessions.

In summary, the findings of this study strongly suggest that the prediction of RW characteristics from observations of VW characteristics and behavior requires the input of multiple predictor variables in order to develop a stable predictive model, and that some of those models are more easily explained in terms of face validity than others. Unfortunately, prediction is not as simple as declaring that a female avatar must be operated by an RW female, or an individual who is highly extraverted in the RW will appear highly extraverted in the VW. With the advancement of both social science and machine learning techniques, researchers in the field should be able to establish the true nature of the relationship between RW and VW identity over time. Each study along the way can be viewed as a stepping stone toward a greater understanding of how to reliably predict RW from VW characteristics.

Acknowledgments We express our thanks to Dr. Celia Pearce, Kathleen Wipf, Jasmine Pettiford, Nic Watson, Hank Whitson, and Patrick Coursey for their work and contributions on this project.

References

1. Bessière, K., Seay, A.F., Kiesler, S.: The ideal elf: Identity exploration in World of Warcraft. Cyber Psychol. Behav. **10**(4), 530–535 (2007)
2. Messinger, P.R., Xin, G., Stroulia, E., Lyons, K., Smirnov, K., Bone, M.: On the relationship between my avatar and myself. J. Virtual Worlds Res. **1**(2), 1–17 (2008)
3. Pearce, C.: Collaboration, creativity and learning in a play community: A study of the University of There. Proc. Digital Games Res. Assoc. (DiGRA), London, UK (2009)
4. Shim, K.J., Pathak, N., Ahmad, M.A., DeLong, C., Borbora, Z., Mahapatra, A., Srivastava, J.: Analyzing human behavior from multiplayer online game logs: A knowledge discovery approach. IEEE Intell. Syst. **26**(1), 85–89 (2011)
5. Guild Wars.: [Online Game], North America: NCSOFT Corporation (2005)
6. Aion.: [Online Game], North America: NCSOFT Corporation (2009)
7. Dunwoody, P., Funke, F.: Testing three three-factor authoritarianism scales. J. Social Political Psychol. submitted for publication
8. Costa, P.T., McCrae, R.R.: NEO PI-R Professional Manual. Odessa, FL: Psychological Assessment Resources, Inc., 1992
9. Stockburger, D.W.: Discriminant function analysis. Multivariate Statistics: Concepts, Models, and Applications [Online]. http://www.psychstat.missouristate.edu/multibook/mlt03.htm. Accessed August, 2013.
10. Brace, N, Kemp, R, Snelgar, R.: SPSS for Psychologists, 4th edn. New York: Routledge (2009)
11. Burns, R., Burns, R.: Discriminant analysis. Business Research Methods and Statistics Using SPSS, pp. 589–608. Thousand Oaks: Sage (2008) (ch. 25)

12. Yee, N.: Gender-bending. The Daedalus project [Online]. http://www.nickyee.com/eqt/genderbend.html#5. Accessed August, 2013.
13. Bergstrom, K., Jenson, J., de Castell, S.: What's 'choice' got to do with it? Avatar selection differences between novice and expert players of World of Warcraft and Rift, pp. 97–104. Proc. Int. Conf. Foundation of Digital Games, Raleigh, NC: ACM Press (2012)
14. Yee, N., Ducheneaut, N., Yao, M., Nelson, L.: Do men heal more when in drag? Conflicting identity cues between user and avatar, pp. 773–776. Proc. 2011 Annu. Conf. Human Factors in Computing Syst. (CHI'11). ACM Press, New York (2011)
15. Yee, N.: The demographics, motivations and derived experiences of users of massively multi-user online graphical environments. PRESENCE: Teleoperators Virtual Environ. **15**, 309–329 (2006)
16. Pearce, C.: The truth about baby boomer gamers. Games Cult. **3**, 1–25 (2008)
17. Judge, T.A., Bono, J.E., Illies, R., Gerhardt, M.W.: Personality and leadership: A qualitative and quantitative review. J. Appl. Psychol. **87**, 765–780 (2002)

Analyzing Effects of Public Communication onto Player Behavior in Massively Multiplayer Online Games

Kiran Lakkaraju, Jeremy Bernstein and Jon Whetzel

Abstract In this preliminary work, we study how public forum communication reflects and shapes virtual world behavior. We find that in-game groups have differential public posting habits; that player behavior is reflected in public communication (in particular, players who attack more are mentioned more in the public forums), and finally that public and personal communications are linked, those who speak together publicly also speak together privately.

1 Introduction

Massively multiplayer online games (MMOGs) are a fruitful domain to study a variety of social and behavioral issues. One interesting question is understanding how a players' behavior in the game reflects, or is shaped by a players' "real world" characteristics. In this work, we focus on how communicative behavior of players interacts with their behaviors within a game.

We cast this exploration through the idea of creating *Richly communicative non-player characters* (RC-NPCs). Socialization is an important motivation for playing massively multiplayer online games (MMOGs) [1]. Many games rely on socialization between players, however in some cases, there may be a lack of players to socialize with, which can be especially true in the initial stages of a game. Current technologies for nonplayer characters (NPC) do not provide rich interaction between players, often relying on prescripted dialog trees.

We suggest the development of agents that have capabilities for rich communication with other players. Unlike traditional NPCs, which focus on providing limited services (such as a shopkeeper) to a player, we envision agents that mimic the behavior of real players. These NPCs will act like a player, trading, fighting, and even communicating with other players and NPCs. We call these agents *richly communicative nonplayer characters* or RC-NPCs for short.

K. Lakkaraju (✉) · J. Bernstein · J. Whetzel
Sandia National Labs, Albuquerque, NM, USA
e-mail: klakkar@sandia.gov

J. Whetzel
e-mail: jhwhetz@sandia.gov

M. A. Ahmad et al. (eds.), *Predicting Real World Behaviors from Virtual World Data*,
Springer Proceedings in Complexity, DOI 10.1007/978-3-319-07142-8_3,

While there have been many studies on the behaviors of players, but less on understanding the communicative behaviors of players. Data have been collected for a few games, but these often focus on the personal communication between players [2]. Another aspect of communication are the "public forums"—places where all players in a game may post news and thoughts which can be commented on by other players[1].

Forums often have lively conversations and are places to discuss opinions (often highly controversial) on other players, game modifications, and activity within the game. Forums are also a place for players to recruit others. Since forums are public and can be viewed by anyone, they are often the first measure of the social interaction in a game. Even though only a small fraction of players post, many people read the posts[2].

We will use forum data as a representation of the "real world" behavior of players. The unconstrained nature of forums, where players can write nearly whatever they want, even if it is not relevant to the game, allows for the explicit and implicit expression of attitudes and sentiment. For instance, there is much work on inferring emotional states and other aspects from subject text [3, 4]. Forum data are also easily gathered, which is another benefit of using it.

In this preliminary chapter, we study the the relationship between communicating publicly and behavior within the games. We will discuss three patterns found within a 2-year data set of game behavior obtained by the authors:

1. In-game groups have differential public posting habits. In particular, the notion of "country" and "race" (which are somewhat tied together), impacts public communication frequency.
2. Public communication reflects in-game behavior. Specifically, we find a correlation between players who attack and players who are mentioned publicly.
3. Public and personal communications are linked. Specifically, those who talk to each other publicly also talk to each other in private.

By studying these relationships, we hope to increase our knowledge of real world/virtual world interaction.

2 Related Work

Learning behavior profiles is the closest work to ours. We can divide these into two categories: those that focus on player demographics and linking to in-game behavior (what we call the "demographic" aspect), and those who categorize players based solely on in-game, often kinetic behavior.

The main methodology in the demographic perspective is the use of surveys and other instruments to ascertain personality traits [5, 6], or motivations for play,

---

[1] For instance, *World of Warcraft* offers a forum at http://us.battle.net/wow/en/forum/.

[2] For instance, in the data set we investigate below, public forums were viewed more than 5 million times, even though the number of poster was a tiny fraction of the people who actually played.

[1, 7] and other factors. These are then linked to behavior in the game, including communication.

This study forms an important part of the puzzle by providing us clues on how people will play. However, they focus on the link between human and in-game. In creating RC-NPCs, we do not necessarily need to know the personality characteristics, but rather the characteristics of play in game. This leads us to behavioral profiling.

Many studies have attempted to identify roles and player types just from "telemetry data" [8]. Drachen and others [8, 9] used unsupervised clustering approaches to find evidence for different types of players with a variety of skills. For instance, players in *BattleField 2: Bad Company 2* were divided into roles like "assault specialist," "medical-engineer," and "sniper" [8].

Other behavioral profiling approaches focused on churn prediction [10].

Most of these works have focused on what we call "kinetic" behaviors—actions that can directly change the state of the game world. This can include combat and trading activities, participating in raids, etc. Only a few have studied communicative behaviors, which is of interest for building bots for socialization. Ducheneaut and Moore [2] studied interaction patterns in *Star Wars Galaxies*.

Our work is different from these in terms of our end goal. Since we want to design RC-NPCs, we need to know how in-game action affects communication behavior.

3 Description of Game X

Game X[3] is a browser-based exploration game which has players acting as adventurers owning a vehicle and traveling a fictional game world. There is no winning in Game X, rather players freely explore the game world and can mine resources, trade, and conduct war. There is the concept of money within Game X, which we refer to as marks. To buy vehicles and travel in the game world, players must gather marks. There is a vibrant market-based economy within Game X.

Players can communicate with each other through in-game personal messages, public forum posts, and in chat rooms. Players can also denote other players as friends or as hostiles. Players can take different actions, such as:

1. Move vehicle
2. Mine resources
3. Buy/sell resources
4. Build vehicles, products, factory outlets
5. Fight nonplayer characters (NPCs)
6. Fight other players.

Players can use resources to build factory outlets and create products that can be sold to other players.

[3] To preserve the confidentiality of the game, we have anonymized some of the terms describing the game. No descriptions of gameplay dynamics have been changed.

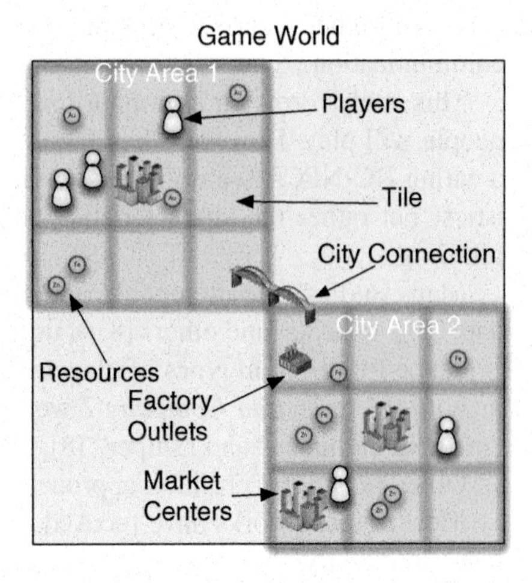

Fig. 1 Overview of the Game X world

Unlike other MMO's like World of Warcraft (WoW) and Everquest (EV), Game X applies a "turn system." Every day each player gets an allotment of "turns." Every action (except communication) requires some number of turns to execute. For instance, if a player wants to move his vehicle by two tiles, this would cost, say, 10 turns. Turns can be considered a form of "energy" that players have.

The use of turns has two major impacts:

1. Players with varying time commitments can play together. Since everyone is limited to the same amount of actions per day, players with minimal time on their hands are not at a disadvantage. In contrast, in WoW, players leveling and experience can depend highly on the amount of time they play (e.g., "grinding").
2. Players have to think about their moves ahead of time. Because there is a limit on turns, players must think and plan ahead before making their moves.

Figure 1 is a schematic depicting the playing space of Game X. Players move from tile to tile in their vehicles. Tiles can contain resources and/or factory outlets and market centers. Only one factory outlet/market center may exist on a tile. The world is 2D, and does not wrap around.

Players can gather resources from tiles and sell them to market centers. Factory outlets allow the creation of new goods from resources—i.e., producing steel from iron ore. More advanced factory outlets exist which can create more advanced objects—i.e., taking steel and producing a sword. Gathering resources and selling to factory outlets is the main way of gaining marks in Game X.

Factory outlets and market centers can be built by players. Creating these structures is relatively straightforward and does not take much in terms of marks or experience. The difficulty lies in maintaining the structures. In order to prosper, the structures require certain resources. Once built, supplying your structures with the necessary resources can be time consuming. Joining a guild (see below) can be helpful as members of the guild can supply your structure.

Table 1 Summary of properties of the groups in Game X

	Nation	Agency	Guild	Race
Number	Fixed, 3	Fixed, 2	Dynamic, many	Fixed
Membership type	Open	Open (req's)	Closed	Open
Modifiable	Yes	Yes	Yes	No

Players can also engage in combat with NPCs, other players, and even market centers and factory outlets. Players can modify their vehicles to include new weaponry and defensive elements. Players have "skills" that can impact their ability to attack/defend.

3.1 Groups in Game X

There are four types of groups a player may belong to. Table 1 summarizes the properties of these.

3.2 Nations

There are three nations a player may join. We label them as A,B, and C. A player may choose not to join a nation as well.

Nations are fixed and defined by the game creators. Nation membership is *open*, players may join any nation they wish at any time and leave at any time.

Joining a nation provides several benefits:

1. Access to restricted, "nation controlled" areas
2. Access to special quests
3. Access to special vehicles and add-ons.

Nations have different strengths; one nation may be better suited for weaponry, and thus has more weaponry related add-ons. Another may be suited for trading.

Completing quests for a nation increases a player *stature* toward the nation which leads to access to special vehicles and add-ons.

Wars occur between nations.

3.3 Agency

An agency can be thought of as a social category. There are two agencies, X and Y. A player can only be a part of one agency at any time. To gain membership to an agency, certain requirements need to be met, but if those are met, anyone can join the agency.

Certain vehicles are open to particular agencies.

3.4 Race

Players may chose their race when they create a character. Different races have strengths in certain areas, implemented as different initial levels of skill. Race is fixed and cannot be changed once chosen. Race also determines starting location.

Race does not seem to play a strong role in the dynamics of the game.

3.5 Guild

Game X also allows the creation of player led guilds. These guilds allow members to cooperate to gain physical and economic control of the game world. Guilds comprise a leader and board who form policy and make decisions that impact the entire guild membership.

Guilds can be created by any player once they have met experience and financial requirements. Guilds have a minimum membership of one, and no upper limit on size.

Apart from the officers, there are the "privileged guild members"—a special set of guild players which is considered important. Finally there are the regular guild members.

Guilds are *closed*—players must submit an application and can be denied membership.

Guild members have access to private communication channels.

Guilds have a "guild account" which can store marks from players (taken in the form of taxes). These marks can be redistributed at the will of the CFO.

3.5.1 Communication in Game X

Game X includes three methods by which players can communicate with each other:

1. Personal messages: An email like system for communicating with other players, or in some cases, groups of players.
2. Public forum: A Usenet like system in which players can post topics and replies (see below).
3. Chat roomt: An IM like system for players to chat with others in their guild.

The structures of the forums are shown in Fig. 2.

Each forum post includes the name of the players who posted an image of their avatar in the game, and their guild affiliation.

3.5.2 Forum Based Measures

We have data on more than 700 days from the game. Included in this dataset are posts from seven different forums within the game. One of the forums is meant for role

Fig. 2 Forum structure in Game X. Each forum can have multiple topics, and each topic can have multiple posts

Table 2 Overview of post/authors/and topics for each forum. *Italic entries* are the max values for the column. These are calculated over the entire 737-day period

Forum	# Posts	# Authors	# Topics	Posts/topic
1. (NRP)	16847	1494	*2468*	6.8
2. (RP)	*62669*	*2244*	1813	34.6
3. (NRP/RP)	35069	1909	1240	28.3
4. (NRP/RP)	9544	1391	223	*42.8*
5. (NRP/RP)	11047	1424	2091	5.3
6. (NRP/RP)	13326	1497	875	15.2
7. (NRP/RP)	1778	341	286	6.2

playing (RP) discussion, that is, all players must discuss in the role of their character. One forum is meant for Nonrole playing discussion (NRP). The rest of the forums are both RP and NRP.

Table 2 provides some high level statistics of the forums. We can see that forum 2, which was only RP, was the most popular in terms of posts. However, the number of topics was low—indicating higher average topic lengths.

Figure 3 shows the posts per day over the entire time period of the dataset. We chose the 50-day time period starting from day 500 as our evaluation period. This time period had a relatively stable rate of posts per day, new players per day, and active players per day. Our goal was to reduce the impact of posting behavior from "newbies." We mean the period from 500 to 550 in the following when we use the term "evaluation period."

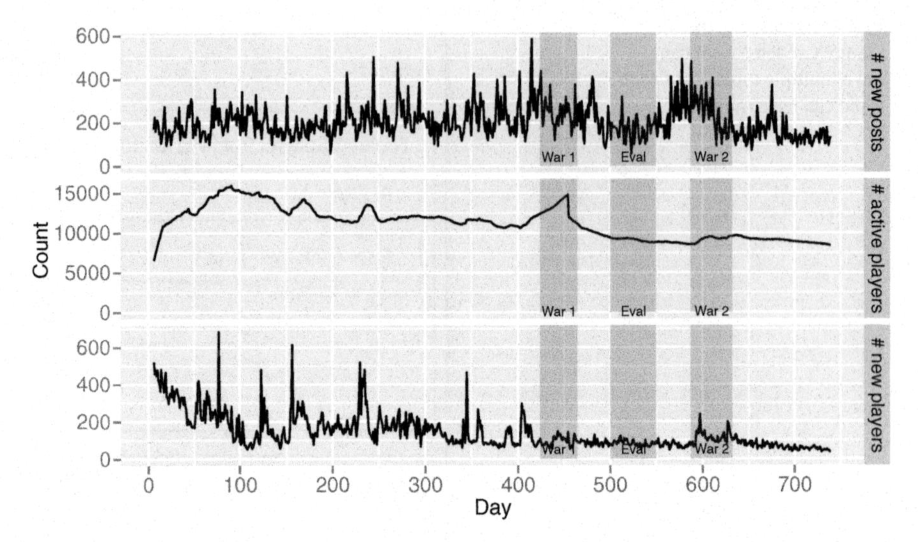

Fig. 3 New posts per day from day 5 to 739. *Highlighted time periods* indicate the two wars and the evaluation period

There are three ways of communicating with others on the forum:

1. Create a new topic.
2. Post on a topic created by another player.
3. Post on a topic and quote another player.

In this work we only consider the second type of public interaction, which we call "coposting."

Definition: Coposters The *coposters* of a player *p* as all players who have posted in a topic that player *p* has also posted in.

In Fig. 2, players p1 and p2 are coposters, as are p0 and p1; and p0 and p2 because they all have posted on the same topic (t0).

We choose to study "coposting" because it directly encodes participation in a conversation and it can be measured in many types of social media.

Players can reference others' posts and players by either quoting them or mentioning them in a post. A players' "mentions" is the number of times he was mentioned in another post.

4 Characteristics of Posters

What motivates players to publicly communicate? One part of the equation is the personality of the player [5]. We hypothesize that other, in-game factors can influence communication behavior as well. This could, for instance, be due to norms among subgroups.

Table 3 Context of posting evaluation period

	Count
# of posts	8543
# of topics	677
# new topics	470
# of posters	750

In this section, we focus on identifying in-game characteristics of players that could influence public communication habits. These include groupings within the game (nation, guild, race, and agency), and the sex and experience of a player.

4.1 Evaluation

Table 3 describes some statistics about the forums during the evaluation period. Figure 3 indicates that this period was relatively stable in terms of number of posts. Even so, there were many active topics and newly created topics. The number of players (750) gives us a good sample of the population. The total number of active players during the evaluation time period was approximately 12,000, so the number of posters was about 6 % of the active player population during the evaluation time period.

Figure 4 shows, per day, several measures:

1. # of posters: The number of players who posted during the day (provided for context).
2. # of posts: The number of posts created that day (provided for context).
3. % Male: Fraction of the players who posted that day that were male.
4. % Guild: Fraction of the players who posted that day that were in a guild.
5. Entropy over race: The entropy over the proportions of players in the different races.
6. Entropy over nation: The entropy over the proportions of players in the different nations.
7. Entropy over agency: The entropy over the proportions of players in the different agencies.

We can see relatively stable behavior across all measures.

The "% male" measure varied between 0.8 and 0.9. This makes sense, as the overwhelming majority of characters in Game X are male.

Percentage of posters who were in a guild showed a 10 % range as well, from 0.875 to 0.975. The high correspondence could be because of a confound—those who have more experience participate in guilds more and they communicate more. To check this, we calculated two variables tracking the total amount of turns a player spent in his lifetime by the end of the evaluation period ("LifetimeTurns") and whether the player posted during the evaluation period ("EvalPosted"). We then binned players based on LifetimeTurns, and calculated the percentage of players who were part of a guild and who posted during evaluation period.

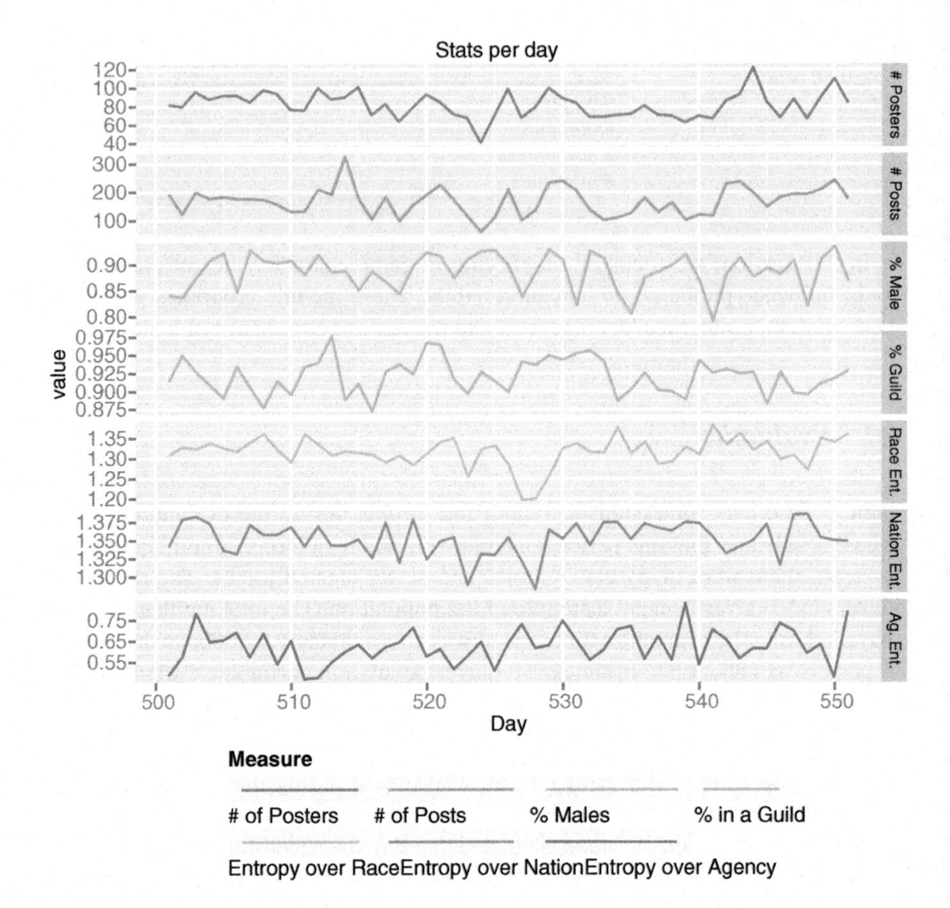

Fig. 4 Per day measures. See text for details

Figure 5 shows the results. As expected, we see that most players with a high "LifetimeTurns" are in guilds, and in addition, they are more likely to post. Thus, we don't know yet whether posting is due to being in a guild or because they are high in turns.

Figure 4 also indicated a relatively uniform distribution over Race and Nation. However, further observation indicated a statistically significant difference in mean proportion over days (Race: ANOVA with $F = 144.7$, $d.f = 3.0$, p value $< 2.2^{-16}$; Nation: ANOVA with $F = 46.86$, $d.f. = 3.0$, p value $< 2.2^{-16}$). It turns out that Nation 1 and Race 4 are disproportionately involved in posting.

The reason for Nation 1 and Race 4 being more involved in public posting is most likely game specific. One explanation could be the aggressiveness of Nation 1 of the two large scale conflicts in our dataset, Nation 1 participated in both. Note that the evaluation period is between the two wars, and thus may reflect discussion of War 1 (by Nation 1), and may foreshadow War 2. Nation 1 may be making threats during this period that will escalate to war. Further work with the text of the public forum posting can help determine that.

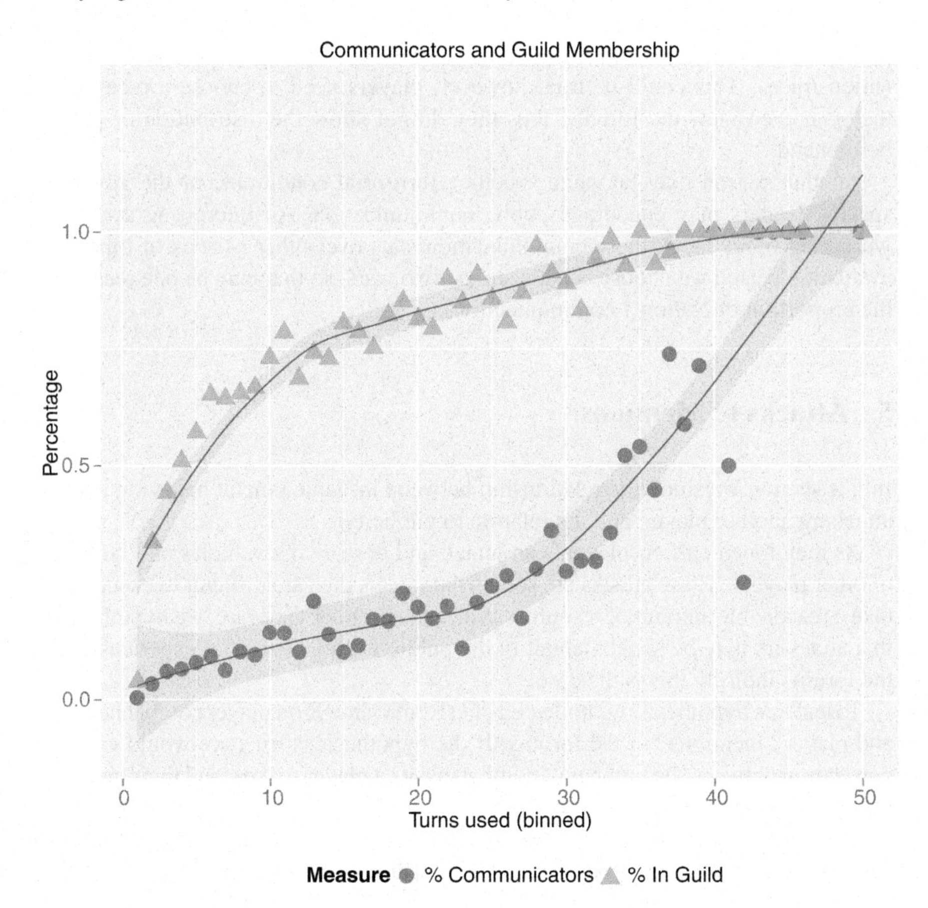

Fig. 5 Communication and guild membership as a function of experience

The difference in race and posting may be tied to nation as well. While any race may join any nation, there are certain normative pairings. It turns out that Race 4 is one of two races that is usually aligned with Nation 1.

Most of the players (mean ≈ 0.8) who posted did not have an agency affiliation.

4.2 Discussion

In general, we can say that public posting is not just a factor of player personality, but could also be influenced by groupings within the game. Future models of player behavior may not need models of personalities, but can assume certain traits based on in-game characteristics.

Further analysis, over other games, is needed to determine the factors that could cause a specific nation/race to be more prevalent. One thought is a selection bias, perhaps individuals who are more prone to communicating choose these specific nations/races. This could be true, however, players need to choose a race before they can even view the forums, thus they do not know the distribution of posters beforehand.

Another reason may be game specific. Territorial conditions, or the advent of specific leaders may encourage public communication. An interesting avenue of work may be to check these communication stats over other periods of time. The evaluation period we choose is between the two wars, so that may be one reason for the proportion of Nation 1 communications.

5 Attacks to Mentions

In this section we study the relationship between in-game kinetic behavior, such as attacking another player, and it's relation to publicity.

As mentioned earlier, players can attack and destroy the vehicles and buildings of other players. These attacks are somewhat rare as they are difficult to execute and take a reasonable amount of resources. Since these attacks are rare, we hypothesized that attackers may be spoken about in the public forums. Personal experience with the forums indicate this is the case.

To test this hypothesis, we collected data from Game X on player combat behavior and player "mentions" in the forums. If the hypothesis is true, we would expect a correlation between the number of combat events a player makes, and their mentions within the forum.

Our data set consisted of 752 players from across the entire data set who had been active during the entire period. We tracked the number of times that each player had:

- attacked another player
- attacked a players' market center
- attacked a players' factory.

We removed all players who did not participate in any combat actions at all.

We also tracked the number of mentions of the player in the public forums over the entire period of time. A confounder in the analysis is any aspect that may also influence a player being mentioned in the forums. Naturally, players who post more will be mentioned more often. In addition, players who have played more will also be mentioned more often. To address this, we segmented the population on turns used and postings according to the 33 and 66 % quantiles and labeled the players as either low/medium/high turns and low/medium/high posting.

To assess the relationship between these two variables, we calculate the Pearson product–moment correlation coefficient between combat activity and mentions. Due to the nonnormal distribution, a rankit[4] transformation was first applied [11]. Table 4

[4] Rankit is the *Rank Inverse Log Transformation* function: $f(x) = \Phi^{-1}((x_r - .5)/n)$, where x_r is the rank of x and Φ^{-1} is the inverse normal cumulative distribution function.

Table 4 Correlations between combat activity and mentions

	Low turns	Medium turns	High turns
Low posting	0.062	0.182	0.023
	(-0.110, 0.230)	(-0.038, 0.385)	(-0.286, 0.329)
Medium posting	*0.239*	*0.344*	*0.277*
	(0.014, 0.441)	(0.147, 0.516)	(0.056, 0.472)
High posting	*0.396*	*0.330*	*0.477*
	(0.092, 0.632)	(0.116, 0.515)	(0.336, 0.597)

shows the Pearson product–moment correlation coefficient (r) and 95 % confidence interval, segmented by the low/medium/high turns and posting values. Bold face entries are significant at 95 %.

5.1 Discussion

The results indicate evidence for a correlation between combat behavior and mentions. This is mediated by the amount of posting one does, players who never post do not have significant correlations. For medium and high posters, however, significant and high correlations occur.

While only content analysis can tell us what is being said about players, we can clearly see a relationship between in-game behavior and posting on the public forum. Note that these are kinetic actions that translate to communicative actions by other players.

The impact of this future behavioral modeling is important to note. RC-NPCs may need to discuss current events and other players in the world.

6 Public and Private Overlap

Our final question is on the relationships between individuals. Will players who interact in the game world also communicate with each other in the public realm? This is an important question for other social media as well.

We addressed this question by comparing the coposting network (described in Sect. 3.5.2 and below) with other "private relationship" networks.

Usenet type discussion boards, due to their similarity to the forums in Game X, will have a notion of coposting. Facebook discussions or comment streams can also be analyzed for coposting behavior. Comment threads (for instance in the social bookmarking/commenting site Reddit) can also be analyzed for coposting behavior.

We constructed the "coposter" network, denoted by G_{cp} by calculating the coposters for every player during the evaluation period of our data set. The nodes in the network are players, and an edge exists between nodes if either of the players is a coposter to the other. The edges are undirected. Each edge is weighted by the number of topics that both players posted on. So a value of 5 would indicate that the two players have both posted on 5 different topics during the evaluation time span. Self edges were removed—thus orphan topics (with no other posts except the original), were not counted.

The degree distribution of G_{cp} (not shown here due to space constraints) indicates many people that only have a single coposter (indicating topics with only two posts). There are several people with an extremely high degree, which was most likely due to participation in topics with many posts that spanned many months.

6.1 Private Relationship Measures

For each edge in the G_{cp} between players p_i and p_j we measured the following variables between the players:

- *Friendship* Are either of the players friends of each other?
- *Hostility* Are either of the players hostile to each other?
- *Personal* Messaging] How many personal messages occurred between p_i and p_j? (threshold of five)
- *Trades* How many trades occurred between p_i and p_j? (threshold of five)
- *Nation* Are the two players part of the same nation?
- *Guild* Are the two players part of the same guild?

We listed two players as *communicating* via personal messages if they sent and/or received more than five messages. We listed two players as *traders* if there have been more than five trades between the players. These thresholds were set in order to remove spurious relationships.

The *link overlap on relationship R* is the percentage of players who are coposters and have the relationship R. For instance, a link overlap of 0.7 on relationship "Friendship" means that 70 % of coposters also were friends.

Recall that several of the topics had hundreds of posts. In these cases, it may be that players were responding to long running topics (such as a feature proposal topic). Coposting on such a topic may not indicate a relationship between players. To address this, we calculate the link overlap by only considering pairs of players who had coposted on several topics—i.e., filtering edges based on edge weights.

More precisely, let $N(x)$ be the set of all edges (pairs of nodes (i, j)) in G_{cp} such that the weight on the edge is greater than or equal to x. The link overlap for coposting threshold x is defined as:

$$L(x) = \frac{1}{|N(x)|} \sum_{i,j \in N(x)} M_R(i, j)$$

where,

$$M_R(i, j) = \begin{cases} 1 & \text{If (i, j) satisfy the relation } R \\ 0 & \text{otherwise} \end{cases}$$

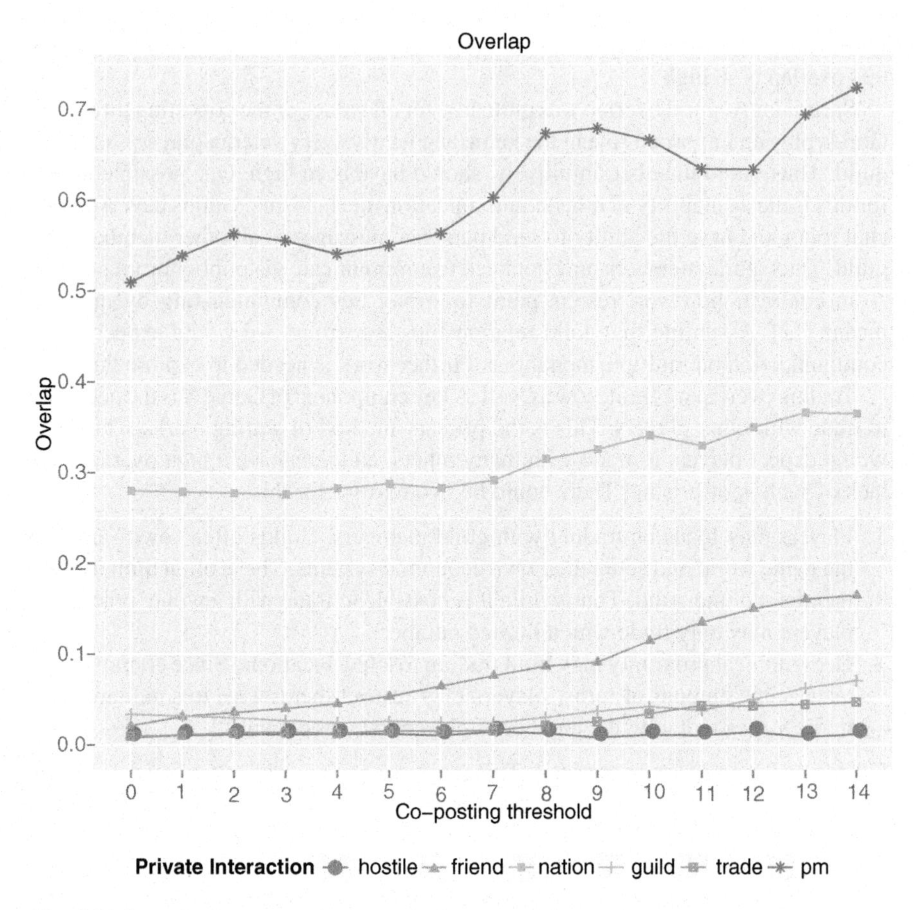

Fig. 6 Link overlap between coposting and private interaction measures

$M_R(i, j)$ represents the private action measure R. For instance, $M_{friendship}$ is 1 if the two players i and j are friends. $M_{communication}$ is 1 if the two players had exchanged more than five messages.

7 Results and Discussion

Figure 6 shows the link overlap for different coposting thresholds.

We can see that nation affiliation had a steady overlap value of approximately 0.255 and peaking at 0.355. Nations do play a role in the large scale conflicts that take place in Game X. The evaluation time period started 40 days after the end of the first war, so it is possible that nation affiliation was still an important attribute. Another factor with nation affiliation: there are only three nations (plus "unaffiliated"). The high overlap value may be just a random effect. Note that we consider two people who are unaffiliated to be in the same nation. Further work will try to identify why this overlap is so high.

Surprisingly, guild overlap was quite low for all values of the coposting threshold. Guilds play an important role in the game and nearly every veteran player is part of a guild. Thus, guild member, intuitively, should have been high. One possible reason for this is the availability of other communication mechanisms. Guilds have a private chat room and have the ability to send personal messages to all other members of a guild. Thus, guild members may not need to communicate via public forum posting.

In contrast, however, results point to strong ties communicating by multiple means [12]. Assuming that guild relationships are strong, we would expect to see communication on multiple modalities. Further work is needed to explore this.

Trading overlap was quite low as well. A key component of Game X is the necessity to trade with other players. This is the primary method of getting marks. Thus, one would expect players to trade with many others, and thus have a high overlap. The lack of such is surprising. There could be two reasons for this:

1. Players may focus on trading with guild members. Guilds often "own" areas of the game world and setup their own economic systems. These often limit trade to members of the guild. Thus, while it is possible to trade with anyone, practically players may only trade with a limited number.
2. Geographical proximity may limit trading to a few locations. Since all movement takes some amount of turns, players may restrict themselves to small areas for trading, thus reducing the number of players they trade with.

Hostile overlap was quite minimal. This indicates players who were listed as hostile to each other did not copost publicly. This makes sense intuitively, if we consider coposting as a measure of the bond between players. However, in some cases, coposting can be used to "troll" others, that is, provide insulting or negative messages. These results show that while that may exist, it does not seem to have a large impact.

Friend overlap was higher than guild, trade and hostile overlap, but experienced change as a function of the coposting threshold, going from close to 0.025 to 0.15. Friendship relationships are of relatively low number, unlike other measures. Thus, they may be more affected by the spurious edges in the coposting network. There could also be a relationship between friendship and coposting, indicating that friends are more likely to copost.

The personal messaging overlap is the most interesting aspect. It is the largest by far, starting at close to 0.5 and peaking at a little higher than 0.7. It seems from this that public and private communications do interrelate.

8 Conclusions, Discussion and Future Work

The development of RC-NPCs can help increase socialization in MMOGs even when there are not many people playing the MMOGs yet. This will lead to a more rewarding experience for the players. To build RC-NPCs, we need an understanding of how in-game behavior relates to in-game communication. We focus on a somewhat neglected area, public communications via in-game forums and identify patterns of behavior and communication that can be used to develop RC-NPCs.

Through an analysis of 2 years worth of data from the MMOG Game X, we have identified the following general patterns:

- Player communicative behavior is influenced by in-game groups.
- Player behavior in-game is affected by his "popularity/notoriety" in the public communication sphere.
- Player relationships in-game are reflected in relationships via public communication channels.

These three patterns can help progress toward agent-based models of communicative agents.

Our future work will focus on content analysis of the posts. An intriguing question is what we call "Expression to Action" (E2A). When does talk *about* action translate into action? Through our data set we can evaluate this relationship.

Acknowledgments Sandia National Laboratories is a multiprogram laboratory managed and operated by Sandia Corporation, a wholly owned subsidiary of Lockheed Martin Corporation, for the US Department of Energys National Nuclear Security Administration under contract DE-AC04-94AL85000.

References

1. Yee, N.: Motivations for play in online games. Cyberpsychol. Behav. **9**(6), 772–775 (2006)
2. Ducheneaut, N., Moore, R.J.: The social side of gaming: a study of interaction patterns in a massively multiplayer online game. Proceedings of the 2004 ACM conference on computer supported cooperative work, ser. CSCW '04, pp. 360–369. ACM, New York (2004) [Online]. http://doi.acm.org/10.1145/1031607.1031667
3. Tausczik, Y.R., Pennebaker, J.W.: The psychological meaning of words: LIWC and computerized text analysis methods. J. Lang. Soc. Psychol. **29**(1), 24–54 (2010) [Online]. http://jls.sagepub.com/content/29/1/24
4. Pang, B., Lee, L.: Opinion mining and sentiment analysis. Found. Trends® Inf. Retr. **2**(1–2), 1–135 (2008) [Online]. http://www.nowpublishers.com/articles/foundations-and-trends-in-information-retrieval/INR-011;jsessionid=5C25DD6BCF066D71D33A9D0EB8FEAF74
5. Yee, N., Ducheneaut, N., Nelson, L., LIkarish, P.: Introverted elves & conscientious gnomes: the expression of personality in world of Warcraft. Proceedings of CHI 2011, Vancouver, (2011)
6. Yee, N., Harris, H., Jabon, M., Bailenson, J.N.: The expression of personality in virtual worlds. Soc. Psychol. Personal. Sci. **2**(1), 5–12 (2011)
7. Yee, N.: The demographics, motivations, and derived experiences of users of massively multi-user online graphical environments. Presence (Camb). **15**(3), 309–329 (2006) [Online]. http://dx.doi.org/10.1162/pres.15.3.309
8. Drachen, A., Sifa, R., Bauckhage, C., Thurau, C.: Guns, swords and data: clustering of player behavior in computer games in the wild. 2012 IEEE conference on Computational Intelligence and Games (CIG), pp. 163–170, Sept. (2012)
9. Drachen, A., Canossa, A., Yannakakis, G.: Player modeling using self-organization in tomb raider: underworld. 2009 IEEE symposium on Computational Intelligence and Games (CIG), pp. 1–8, Sept. (2009)

10. Borbora, Z., Srivastava, J.: User behavior modelling approach for churn prediction in online games. Privacy, Security, Risk and Trust (PASSAT), 2012 international conference on Social Computing (SocialCom), pp. 51–60, Sept. (2012)
11. Bishara, A.J., Hittner, J.B.: Testing the significance of a correlation with nonnormal data: comparison of pearson, spearman, transformation, and resampling approaches. Psychol. Methods. **17**(3), 399–417 (2012) (pMID: 22563845)
12. Haythornthwaite, C.: Strong, weak, and latent ties and the impact of new media. Inf. Soc. **18**(5), 385–401 (2002) [Online]. http://www.eric.ed.gov/ERICWebPortal/detail?accno=EJ668329

Identifying User Demographic Traits Through Virtual-World Language Use

Aaron Lawson and John Murray

Abstract The paper presents approaches for identifying real-world demographic attributes based on language use in the virtual world. We apply features developed from the classic literature on sociolinguistics and sound symbolism to data collected from virtual-world chat and avatar naming to determine participants' age and gender. We also examine participants' use of avatar names across virtual worlds and how these names are employed to project a consistent identity across environments, which we call "traveling characteristics."

1 Introduction

This chapter presents results from VERUS, a large-scale study of the relationship between virtual-world (VW)/online behavior and real-world (RW) characteristics. The program goal was automatically predicting major RW demographic attributes using only VW behavior. These RW attributes include age group, gender, ethnicity, income level, education level, leadership role, and urban/rural background among others. More than one thousand participants volunteered RW demographic information about themselves and allowed the recording of their online behaviors, including text chat data produced during online activities and names chosen for online personae (avatars). SRI gathered hypotheses from the theoretical sociolinguistics literature, phonology and sound symbolism, semantics, and discourse analysis and made empirical observations to generate features for determining RW attributes that could be combined in a global model using statistical classifiers.

Some of the major observable features had application to several subareas, such as factors that related to neophyte online behavior (using RW names and written language syntax). In Table 1, we give examples of those features that correlated highly with the prediction of two conditions: age and gender.

A. Lawson (✉)
Speech Technology and Research (STAR) Lab, SRI International, Menlo Park, CA, USA
e-mail: aaron.lawson@sri.com

J. Murray
Computer Science Laboratory, SRI International, Menlo Park, CA, USA
e-mail: john.murray@sri.com

M. A. Ahmad et al. (eds.), *Predicting Real World Behaviors from Virtual World Data,*
Springer Proceedings in Complexity, DOI 10.1007/978-3-319-07142-8_4,

Table 1 Prominent chat features

Behavior	Class	Example
Hedging/Uncertainty	Female	Questions, "I don't know"
Command forms	Male	"Heal me!"
Use of slurs	Male	"You homo!"
Direct apologies	Female	"I'm sorry"
Indirect apologies	Male	"Ooops," "my bad"
Empathy	Female	"I like X," "u OK?"
Use of modal verbs	Female	can, could, would, should, etc.
Use of all caps	Youth	"STOP BEING DUMB"
Frequent use of ellipsis	Adult	"if you bring up your questlog. . . "
Using commas, apostrophes	Adult	"we're done. let's turn in"
Lowercase "i" for "I" and "u" for "you"	Youth	"u losted to 4 pokemno"
Single word utterances	Youth	"come," "yo"

Various windows into RW identity were provided by chat behavior, including word-choice factors such as swearing, insults, apologies, and expressions of certainty; and grammatical factors including verb forms and use of questions. Length of utterance, typographical features, spelling, part of speech, and semantic affect were also found to contribute to identity determination and to apply outside chat data [1]. Avatar-naming conventions in terms of typography (capitalization, special characters, spaces, etc.), phone class of characters used, syllable structure, name endings, use of numbers, and other factors enabled good prediction of both age group and gender [2]. Further, even when online avatar gender differed from RW gender, naming conventions still enabled the correct prediction of RW gender. Participants who used multiple online names exhibited behavior that enabled linking of different names across sessions [3].

The following sections delve more deeply into using language-based features to enable the detection of demographic categories. We focus on age and gender, but also discuss how avatar names relate to demographics and how these names are used across VW environments.

2 Data

We obtained data for this study from the VERUS internal corpus of VW chat and avatar demographics information from the Sherwood and Guardian Academy worlds datasets only. World of Warcraft (WoW) and Second Life data were used for hypothesis generation. The avatar names used in this study were chosen by participants when setting up their character information only for games on the VERUS server environment. Note that no actual avatar names are presented in this paper due to privacy concerns on the part of participants. Table 2 summarizes the total amount of data collected for the project.

Table 2 Data distribution of virtual-world chat in the VERUS project

Game	Turns	Talkers	Tokens
Guardian Academy	914	57	2,688
Sherwood	13,149	271	57,843
Second Life	79	4	392
WoW	2,337	117	56,036
Total	11,214	445	89,521

3 Gender

This section explores gender differences in VW language use. We test the relevance of traditional sociolinguistic observations of males and females in face-to-face conversation to the contemporary space of VW chat interactions in online gaming and collaborative environments. In addition, we study the relationship between a player's RW gender and naming decisions for online personas, or avatars, in the light of linguistic observations based on sound symbolism and naming conventions. The approach taken in this study focused on applying sociolinguistic claims and observations to develop discourse features for characterizing gender in VW chat and looking at other linguistic factors, such as choice of avatar name, to detect gender trends. To expedite the development of features, we examine the rich empirical claims of the sociolinguistic literature to identify known factors that have tended to correlate with male or female speech, especially [4–8]. A primary goal of this study is determining whether these findings apply in the physically distant universe of VW interactions.

3.1 Sociolinguistic Features

Based on the literature, ten features were evaluated: (1) expressions of uncertainty, (2) strong swears, (3) light swears, (4) insults, (5) slurs, (6) questions, (7) modals verbs, (8) expressions of empathy, (9) strong apologies, and (10) indirect apologies. Rules for extracting each of these features were developed and the results were calculated to verify whether gender-based biases existed in the distribution.

Analysis of the results (Fig. 1) is given as *lift*, the degree to which the feature rises above random. Results show that many RW sociolinguistic claims about gender and discourse also hold true in the VW: women were much more likely to use modal verbs, ask questions, use expressions of uncertainty, and use strong apologies than were males. Males were much more likely to use strong swears, slurs, and indirect apologies. The results also demonstrate that some of the categories suggested in the sociolinguistic studies may be too coarse. For example, women are claimed to apologize more frequently than men. However, analyzing the kinds of apologies that occur makes it clear that direct apologies (e.g., "I'm sorry") are more typical of female players, while indirect apologies ("oooops!" or "my bad!") are more associated with males. Similarly, with swears, a breakdown between the kinds of words used existed: light swears were associated more with females and strong swears, more

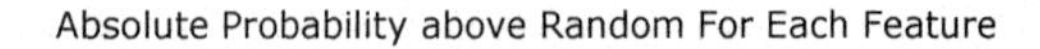

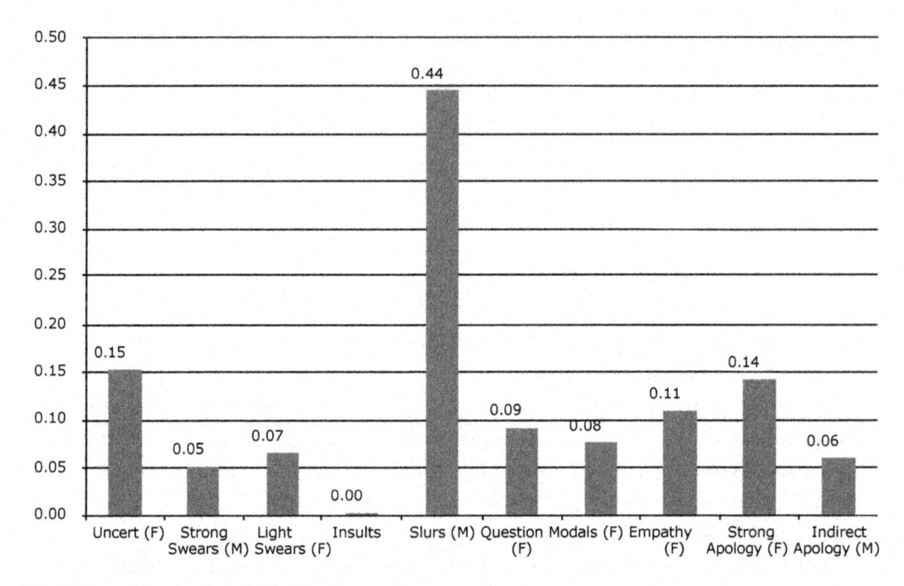

Fig. 1 Prominence of each feature for gender

with males. This observation is actually in keeping with the observations that men are more comfortable with profanity than females, as many of the "light swears" represent approaches to avoid offensive cursing. Slurs were the category most strongly associated with males, and most slurs were homophobic in nature.

3.2 Gender Differences for Emoticon and Ellipsis Use in Chat

Other groups have reported significant gender effects for the use of both emoticons and the ellipsis in online chat. The major hypotheses claim: (1) women tend to use ellipsis more often than men, (2) women tend to use emoticons more often than men, but (3) men tend to use lewd (:P, :D, xD) emoticons more often. To determine whether these claims hold with the data collected by the VERUS team, 9,373 lines of in-game chat were process for these phenomena. Of this data, 5,149 turns were produced by males and 4,226 produced by women. These data came from the Guardian Academy and Sherwood.

The full ellipsis (. . .) frequency was calculated for both genders, with 152 occurrences for males and 116 for females. After adjusting for differences in the priors for both classes, the probability of the ellipsis being associated with females is 48.2 %. The partial ellipsis (. .) was also evaluated, with 208 occurrences for males and 163 for females. The probability of partial ellipsis being used by women was 48.8 %. The distribution of all emoticons across genders was calculated, with 124 instances

Table 3 Avatar rules

Rule no.	Gender affected	Formulation
1	Female	ends in "a"
2	Male	ends in back vowel
3	Female	ends in "y"
4	Male	ends in "er"
5	Male	ends in back or alveolar stop
6	Male	ends in any consonant
7	Male	ends with fricative consonant
8	Male	begins with capital letter
9	Male	contains "x" or "z"
10	Female	contains palatal fricative ("sh")
11	Male	contains a title of nobility
12	Male	is a male census name
13	Female	is a female census name

for males and 164 for females. After adjusting for differences in the priors for both classes, the probability of emoticons being associated with females is 61.2 %. For non-lewd emoticons, males had 36 instances and females, 64 instances. The probability of females using non-lewd emoticons was 68.4 %. The frequency of lewd emoticons was calculated for both genders: males had 57 instances and females, 50 instances. After adjusting for the differences in the priors, the probability of a male using lewd emoticons was 53.3 %.

The conclusions were mixed. No gender effect was seen for either type of ellipsis in this VW data, with lift being less than 3 %. (We consider significant any lift of greater than 5 %.) Emoticons in general and non-lewd emoticons in particular were significant for females, with lifts of 34.0 and 36.8 %, respectively. Lewd emoticons were not significant for males, with lift of 3.3 %. A clear trend exists that female participants tended to use emoticons more often than did males. This use may be a function of hedging or qualifying a turn with an emoticon as a way of softening the potential impact of a statement or ensuring that a turn made in jest was not interpreted as a insult or slight.

3.3 Avatar Names and Gender

For investigating the relationship between gender and avatar naming, 13 rules, largely based on observations from the sound symbolism research (especially [9, 10, and11]), were developed: four for females and nine for males. These included phonetic rules such as female names ending in low vowels; male names ending in back vowels; male names containing "z" or "x"; and female names containing "sh." The association between female names and final low vowels comes from the frequency of female grammatical endings in both Semitic and Indo-European languages. In addition, we included more basic rules, such as the use of female names for female players and male names for male players, based on 2010 US census data. These rules are listed in Table 3.

Table 4 Proposed age rules

Rule no.	Age characterized	Formulation
1	1	contains a number
2	2, 3	is a census name
3	2	contains space or separator in name

Table 5 Age results

Rule no.	AvePrec.	Prec.	Recall	F-score
1	0.82	0.76	0.26	0.39
2	0.76	0.89	0.32	0.47
3	0.62	0.87	0.07	0.13

The highest precision sound-based rules deal with word endings, with those words ending in a fricative consonant being strongly male, and words ending in the central vowel schwa (represented orthographically with "a") being strongly female. Applying the same rules to avatar names from individuals whose RW and VW gender were different enabled the detection of RW gender at a similar high rate of accuracy (> 0.7), despite the mismatched gender. This result was surprising and may show that avatar gender was not playing a significant role in the players' online personas.

4 Age

In this section, we investigate the relationship between the choice of avatar names, chat, and participant age. A total of 305 participants (84 group 1, 178 group 2, 43 group 3) from collects spanning Guardian Academy and Sherwood were used in this section's analysis. Data were collected by matching avatar choice and chat within gaming sessions to demographic information collected in the project. The initial rules were developed based on an examination of the patterns in the avatar names. The notion was that younger avatars would be more innovative and more likely to break the conventions of standard naming, while older users would create names that conformed more to traditional naming conventions.

Three rules were proposed, which are defined in Table 4. The first rule looks at whether a number is used anywhere in the name. Rule two checks whether an avatar name is also a name listed in the US Census report of the 1,000 most common names for 2010. Rule three checks whether a name is divided into a "first name" and "last name" form, which one would associate with canonical names.

4.1 Age Results

These rules were tested against the 305 avatar names with age information from Guardian Academy and Sherwood. The results in Table 5 show that two rules achieved

Table 6 Rule combination

Rule	Class	*P* (%)	*r* (%)
Avatar name is all caps + low use of articles	Youth	92	12
Frequent use of articles but low use of pronouns	Young adult	80	20
Frequent use of both articles and qualifiers/tentative language	Adults	83	14

program goals (2 and 3), reaching precision levels of greater than 0.85. However, if one normalizes for distribution prior probabilities, rule 1 is actually superior to both rules 2 and 3, as far fewer examples of class 1 exist than that of class 2.

4.2 Rule Combination of Names and Chat for Age

The rule combination results are shown in Table 6. These results take the best performing features from both chat and naming and combine the results by using machine-learning techniques. Very young participants tend to use a "telegraphic" chat style with low use of articles and other grammatical particles; low use of overt pronouns (when the meaning is understandable); and frequent use of "shouting" (all CAPS) in both names and chat. Adults have a chat style that conforms much more closely to standard written language with use of overt pronouns, qualifiers, and articles; and use of the flourishes of written language, such as the ellipsis, etc.

5 Traveling Characteristics

Avatars are the primary means by which players navigate with and interact in most massively multiplayer online games (MMOGs). In recent years, both quantitative and qualitative studies have drawn attention to the opportunities for, and practices around, MMOG avatar creation and customization. Quantitative studies have explored player practices around avatar "gender-swapping," class choice, and comparisons of avatar customization options across different games. Qualitative (typically ethnographic) accounts of avatar customization have addressed the relationships and affiliations generated between players and their avatars (usually focusing on one avatar and in one MMOG). Celia Pearce's ethnographic look [12] at how a disenfranchised community of There.com users recreated the world within Second Life represents one of the few studies that examine players' transition from one VW to another. In doing so, Pearce opens up for consideration the ways that players create a sense of stability across experiences that are otherwise contingent upon specific games. Our hypothesis is that naming practices represent one primary (and underexplored) means through which players maintain continuity and a stable identity across their MMOG play. To explore this, we draw on more than 1,700 avatar names, from more than 400 players, obtained through a mixed-methods, multisite study of players across

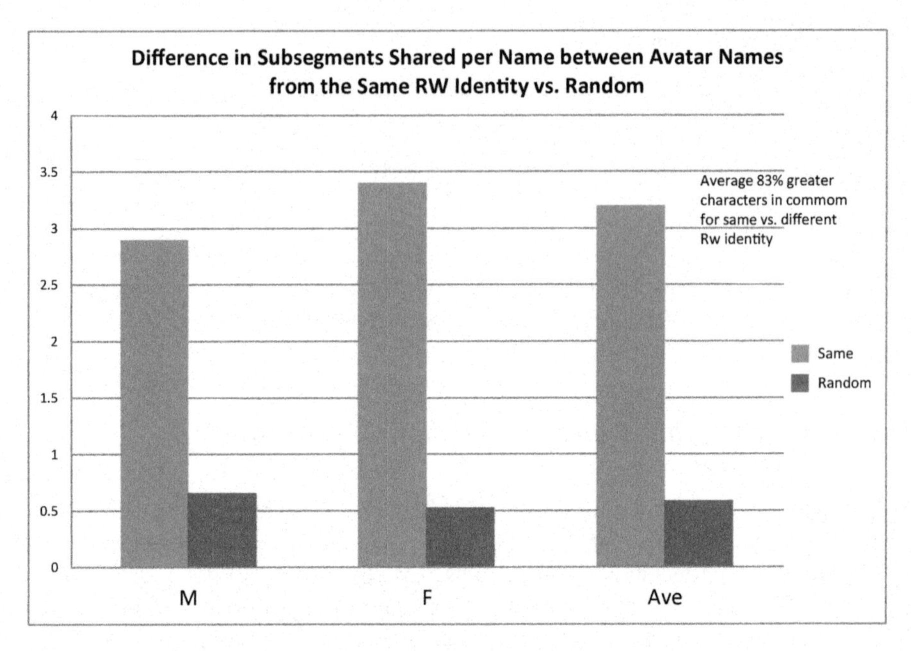

Fig. 2 Reuse of name components across virtual worlds (VWs)

multiple MMOGs. Here, we report on the avatar-naming practices that we identified through both quantitative and qualitative analyses of this dataset.

The quantitative analysis consisted of unigram and bigram similarity measures, and Levenshtein or "Minimum Edit" distance metrics. The goal was to determine the extent to which players share either parts of their avatars across VWs, measured by bigram similarity and Levenshtein distance, or characters from their avatar names, measured by unigram similarity. The analysis compared all of a participant's own avatar names with all other players' avatar names to determine the cohesion of a players' avatar names across worlds. The results showed that considerable "traveling" of names occurs across VWs, with same-player bigrams per name averaging 3.4 for males and 2.9 for females, compared with 0.53 and 0.66 bigrams per name for different player names (see Fig. 2). The Levenshtein distance was 17 % greater for different player names than for same player names, and unigram similarity was 25 % greater for same-player than different. These results demonstrate that players maintain a great deal of measurable continuity across VWs using avatar naming.

Supplementing and extending this quantitative analysis, we also conducted a qualitative exploration and "open coding" of avatar names, identifying significant and widespread instances of naming practices that fall outside those quantitative measures. Specifically, we saw numerous examples of players' intertextual and thematic practices in "branding" multiple avatars. This discussion contributes both conceptual and methodological innovations to the field of MMOG studies. Conceptually, we can identify one significant "traveling characteristic," the means that players use to create a stable identity as they move across avatars and VWs. Methodologically, we can show how quantitative and qualitative analyses of the same dataset can, when developed in tandem, provide a robust picture of an underexplored facet of player practice in MMOGs.

6 Future Research

In this chapter, we show that linguistic features offer high predictive value for identifying user demographics in the VW. We summarize our findings in Fig. 3, with examples of representative chat and naming phenomena and of how they relate to demographics.

Despite the enormous amount of communication occurring online every day, research on VW and online language use is still in its infancy. Three areas offer significant potential to expand our capabilities to identify RW characteristics from VW language usage:

- Semantic-level analysis (word meanings in context) and syntax
- Association of participants through RW cultural references
- Discourse and participant interaction

Semantic analysis is clearly a gap in current approaches, and a pilot study into the kinds of shared RW cultural references in both naming and chat reveals that this sort of information often provides a more focused and fine-grained picture of group association than lexical or typographical features do. In Fig. 3, we see how linguistic features such as typography, syntax, and phonology enable the identifying RW traits such as gender and age. Semantic and cultural references are crucial for understanding more specific multivalent characteristics such as socio-economic status and leadership roles. Semantic analysis also facilitates the effective clustering of individuals based on shared areas of interest and topic (for example, fantasy novels) to enable the linking of individuals who fit a certain profile (individuals who are interested in identified area X are also interested in Y).

Discourse factors—including interpersonal interaction, emotional response to others, nonnative language use, intentionality assessment, and other areas—enable the prediction of RW demographics by combining low-level features with the larger semantic context and permit identity attribution via shared stylistic factors. For example, the discourse acts of greeting (especially being the first greeter) and addressing someone directly by name are associated with high RW leadership scores. Though these "higher-level" features may prove difficult to extract automatically, they offer the potential for new insights into users' online and VW behavior, with many possible applications to related settings, such as artificial-reality environments and games.

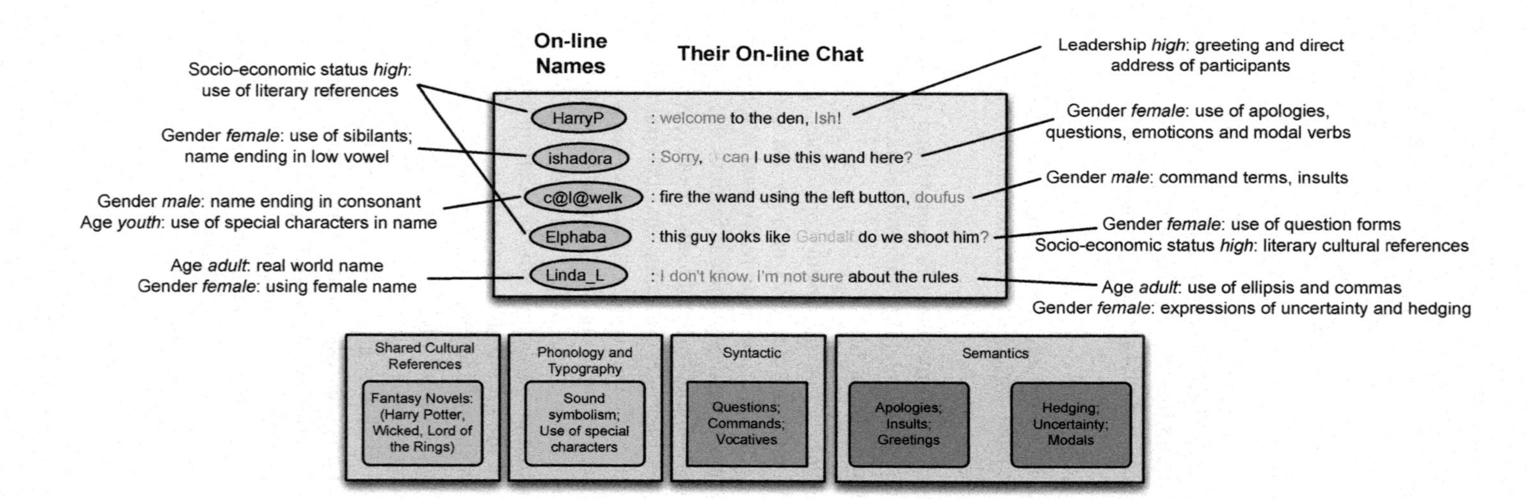

Fig. 3 Summary of chat and naming findings

Acknowledgments The authors acknowledge the Air Force Research Laboratory at Wright Patterson Air Force Base for sponsorship of this research under contract FA8650-10-C-7009.

References

1. Wang, W., et al.: Automatic Detection of Speaker Attributes Based on Utterance Text. Interspeech 2011, Florence, Italy (October 2011)
2. Lawson, A., et al.: Sociolinguistic Factors and Gender Mapping Across Real and Virtual World Cultures. 2nd International Conference on Cross-Cultural Decision Making, San Francisco, CA (July 2012)
3. Lawson, A., Taylor, N.: The Names People Play: Exploring MMOG Players' Avatar Naming Conventions. Canadian Games Studies Association Symposium (May 2012)
4. Lakoff, R.T.: Language and Woman's Place. Harper & Row, New York (1975)
5. Tannen, D.: Gender and Discourse. Oxford University Press, New York (1994)
6. Tannen, D.: Conversational Style: Analyzing Talk among Friends. Ablex, Norwood (1984)
7. Herring, S.C., Paolillo, J.C.: Gender and genre variation in weblogs. J. Socioling.**10**(4):439–459 (2006)
8. Herring, S.: Gender Differences in Computer-Mediated Communication: Bringing Familiar Baggage to the New Frontier. American Library Association Annual Convention, Miami (1994)
9. Gordon, M., Heath, J.: Sex, sound symbolism, and sociolinguistics. Curr. Anthropol.**39**(4, August/October):421–449 (1998)
10. Jespersen, O.: Language: Its Nature, Development and Origin. Allen and Unwin, London (1922)
11. Ohala, J., Hinton, L., Nichols, J.: Sound Symbolism. Cambridge University Press, New York (1994)
12. Pearce, C.: Communities of Play: Emergent Cultures in Multiplayer Games and Virtual Worlds. MIT Press, Cambridge (2009)

Predicting MMO Player Gender from In-Game Attributes Using Machine Learning Models

Tracy Kennedy, Rabindra (Robby) Ratan, Komal Kapoor, Nishith Pathak, Dmitri Williams and Jaideep Srivastava

Abstract What in-game attributes predict players' offline gender? Our research addresses this question using behavioral logs of over 4,000 EverQuest II players. The analysis compares four variable sets with multiple combinations of character types (avatar characteristics or gameplay behaviors; primary or nonprimary character), three server types within the game (roleplaying, player-vs-player, and player-vs-environment), and three types of predictive machine learning models (JRip, J48, and Random Tree). Overall, the most highly predictive, interpretable model has an f-measure of 0.94 and suggests the primary character gender and number of male and female characters a player has provide the most prediction value, with players choosing characters to match their own gender. The results also suggest that female players craft, scribe recipes, and harvest items more than male players. While the strength of these findings varies by server type, they are generally consistent with previous research and suggest that players tend to play in ways that are consistent with their offline identities.

T. Kennedy (✉)
Department of Communication, Popular Culture and Film, Brock University,
500 Glenridge Ave., St. Catharines, ON L2 S 3A1, Canada
e-mail: tkennedy@netwomen.ca

R. (Robby) Ratan
Department of Telecommunication, Information Studies and Media,
Michigan State University, 404 Wilson Road, Rm 409,
East Lansing, MI 48824-1212, USA
e-mail: rar@msu.edu

K. Kapoor · N. Pathak · J. Srivastava
Department of Computer Science and Engineering, University of Minnesota,
117 Pleasant Street S.E., Minneapolis, MN 55455, USA
e-mail: kapoo031@umn.edu

N. Pathak
e-mail: npathak@cs.umn.edu

J. Srivastava
e-mail: srivasta@cs.umn.edu

D. Williams
Annenberg School for Communication and Journalism, University of Southern California,
North University Park, KER 312, M/C 0281, Los Angeles, CA 90089-0281, USA
e-mail: dcwillia@usc.edu

M. A. Ahmad et al. (eds.), *Predicting Real World Behaviors from Virtual World Data*,
Springer Proceedings in Complexity, DOI 10.1007/978-3-319-07142-8_5,

1 Introduction

In many online games, players use characters that are gendered as male or female, but do not necessarily match the players' gender. The present research examines whether it is possible to predict a player's gender from observable gameplay behaviors and characteristics. Previous research on this topic suggests that online game players generally use characters that match their offline gender identity [1, 2], although a large portion of people have experimented at least once with using an oppositely gendered online identity [3]. The choice of character gender is understood to be motivated by reasons related to identity, e.g., to express or explore a role within the game [4], or instrumental reasons related to the gameplay experience, e.g., to influence in-game social interactions or to provide a pleasant aesthetic in the game [5, 6].

While early research focused more on identity-relevant reasons of character gender choice, recent research has provided insights into the instrumental uses of characters by comparing in-game behavioral data with character gender and player gender. Huh and Williams [1] found that male players use oppositely gendered characters more often than females do, but females who use male characters tend to play the game in more stereotypically male ways, e.g., by engaging in player versus player (PvP) combat more and chatting less. This suggests that males are more likely to choose character gender for instrumental reasons, while females choose character gender to express a feminine identity or explore a masculine one. However, Yee et al. [2] found that regardless of player gender, players who used female characters engaged in more stereotypically feminine behaviors (i.e., healing) and less stereotypically masculine behaviors (i.e., PvP combat) than those who used male characters. A reason for this discrepancy may be that these two studies examined two different massively multiplayer online (MMO) games that motivated players to choose character gender in different ways. For example, a game that encourages more PvP combat may induce more instrumental motivations of character gender choice than one that encourages roleplaying.

The present research builds on this previous work by systematically examining three server types within the same online game [roleplaying (RP), PvP, and player-versus-environment (PvE)], allowing for a comparison of play styles and character gender choice motivations encouraged by each type. Further, this research segments the game-based data into two types, avatar characteristics and gameplay behaviors, also providing insights into identity-relevant versus instrumental motivations of character gender choice. Lastly, this research utilizes a novel methodological approach from previous research on this topic. Namely, the data are analyzed with three predictive machine learning models (JRip, J48, and Random Tree), and results are presented along with a set of interpretable statements about the relationships discovered, illustrating a successful convergence of computer science- and social science-oriented approaches.

In summary, this research is guided by one fundamental research question: what in-game attributes predict players' gender? The exploration of this question is segmented into comparisons between three factors: (1) avatar characteristics versus gameplay behaviors; (2) RP, PvP, and PvE server types; and (3) JRip, J48, and Random Tree machine learning models.

2 Dataset

The dataset used for this analysis, provided by Sony Online Entertainment, was from the MMO game EverQuest II (EQII). This game was one of the most popular in the fantasy roleplaying genre during the time window included in the behavioral logs, January 1, 2006 through September 17, 2006. The four terabyte dataset included behavioral logs of over 4,000 players' character attributes and gameplay statistics (described below) matched to player responses to self-report surveys, which participants completed in exchange for a special in-game item provided by the game publisher. There were numerous log and survey measures recorded that are not part of the present analysis. All data recorded were anonymous and privacy protected.

The ground truth of player gender was established through a comparison of survey responses to the question "what is your gender," to which the participants could respond "male" or "female" (the only survey-based measure used in this analysis), and the reported gender provided when the participants opened their EQII accounts, which was included in the behavioral logs. All cases with discrepancies between these two measures were excluded. The gender composition of the sample reflects typical MMO demographics: 83 % male and 17 % female.

2.1 *Three Server Types*

We tested all of the models on three different servers: Antonia Bayle–RP, Nagafen–PvP and Guk–PvE. Each of these servers reflects different kinds of virtual environments within the MMO. For example, players in PvP servers may be expected to engage in less exploration of the environment because they are battling with other players, compared to a PvE server where they cannot battle with other players by default. Interpretations of the substantive differences in play style between these server types with respect to the current analysis are presented in the Discussion section.

2.2 *Four Variable Sets*

We conducted the predictive analyses on four different variable sets, each with a different set of variables (described below) used in the analysis. All variable sets included players who reported being over the age of 18. Also for all variable sets, given the greater proportion of male players, we used supervised oversampling of the female players to resample the data and obtain an overall uniform distribution. Variable Set 1 included participants from all three servers (see Table 1), with gameplay statistics derived from their primary characters (i.e., the character with the most number of hours played), although total number of male/female characters was included in this variable set as well. The sample for Variable Sets 2–4 includes survey respondents from all three servers, with gameplay statistics from all of the characters they play (see Tables 2 and 3), unlike in Variable Set 1. Note: there is a ten-character cap in EQII.

Table 1 Player gender distribution across servers Variable Set 1

	AB RP		Guk PvE		NF PvP	
	Orig	Resample	Orig	Resample	Orig	Resample
Male	1,083	1,107	860	790	276	229
Female	350	1,042	232	848	47	255
Total	1,433	2,149	1,092	1,638	323	484

Table 2 Player gender across servers Variable Sets 2–4

	Males	Females	Total
Antonia Bayle (RP)	5,833	2,115	7,948
	73 %	27 %	100 %
Nagafen (PvP)	3,735	623	4,358
	86 %	14 %	100 %
Guk (PvE)	4,680	1,396	6,076
	77 %	23 %	100 %

Table 3 Player gender and character count resample Variable Sets 2–4

	Original	Resample
Nagafen (PvP)		
Male players	3,735	2,162
Female players	623	2,196
Total		4,358
Guk (PvE)		
Male players	4,680	3,047
Female players	1,396	3,029
Total		6,076
Antonia Bayle (RP)		
Male players	5,833	3,975
Female players	2,115	3,973
Total		7,948

2.3 Variables Used

All of the variables used in the analyses were considered to be observable in the sense that they were public to players of the game or through official websites about the game. This restriction increases the generalizability of our results, i.e., increases the likelihood that someone could utilize the predictive rule statements generated by the analyses to actually predict another player's gender.

The variable sets contained avatar characteristics only, gameplay behaviors only, or a combination of both. This allowed us to ascertain the kinds of variables and combinations that offer the most prediction value and the best interpretability.

The specific variables were chosen based on previous research that describes how male and female players make different avatar choices and play MMOs differently. For example, Yee's [6] research suggests that women are socializers and explorers, whereas men are competitive players. Variable Set 1 includes both avatar characteristics and gameplay variables (Table 4), but only for primary characters. Variable Sets

Table 4 Gameplay attributes in Variable Set 1

Gender of primary character
Race of primary character
Class of primary character
Total quests primary character completed
Total achievement points for primary character
Total experience points for primary character
Total deaths for primary character
Total number of male characters played
Total number of female characters played

Table 5 Gameplay attributes in Variable Sets 2–4

Character gender (Variable Sets 2 and 4)
Character race (Variable Sets 2 and 4)
Character class (Variable Sets 2 and 4)
Total items crafted (Variable Sets 3 and 4)
Total recipes known (Variable Sets 3 and 4)
Total rare items harvested (Variable Sets 3 and 4)
Total quests completed (Variable Sets 3 and 4)

2–4 examine attributes for all characters played (Table 5). Variable Set 2 includes only avatar characteristics (character gender, character race, and character class). Variable Set 3 includes only gameplay variables, but with different variables that offer more focused gameplay (crafting, harvesting, and scribing recipes) than total experience or achievement points might reveal (as in Variable Set 1). Variable Set 4 combines both sets of avatar characteristics and gameplay variables for all characters played.

3 Analyses and Results

Machine learning analyses were conducted using Weka [7], which provided information gain for each variable in the models. We focused on three types of models—JRip, J48, and Random Tree, each of which offers different strengths to the analyses [8, 9]. Random Tree typically offers higher prediction scores because it details the if/then situations in the data, but the rulesets are difficult to interpret. J48 is usually a bit lower, and offers a pruned tree that is more condensed and interpretable than Random Tree output. JRip typically offers the most condensed and interpretable statements in the form of rulesets, but often these prediction values are lower than other algorithms for this very reason. The use of these three models allowed us to present the highest overall prediction value (f-scores), often with Random Tree, as well as the best interpretable scores, through either J48 or JRip.

Each model is presented with a set of performance metrics. Information gain denotes the predictive power of each variable in the model relative to all variables in the model. Precision (P) is the ratio of the number of correctly classified instances over the total number of instances classified as belonging to that class. Recall (R) is the ratio of the number of classified instances over the total number of instances belonging

Table 6 Prediction Variable Set 1 variable information gain

	AB RP	Guk PvE	NF PvP
Gender of primary character	*0.477*	*0.468*	*0.23*
Race of primary character	0.088	0.153	0.085
Class of primary character	0.062	0.06	0.207
Total quests primary character completed	0.024	0.029	0.143
Total achievement points for primary character	0.013	0.025	0
Total experience points for primary character	0	0	0.156
Total deaths for primary character	0	0	0
Total number of male characters played	*0.514*	*0.521*	*0.287*
Total number of female characters played	*0.484*	*0.517*	*0.36*

Table 7 Prediction Variable Set 1 primary characters

	JRip			J48			RT		
AB (RP)	*P*	*R*	*F*	*P*	*R*	*F*	*P*	*R*	*F*
Males	0.96	0.89	0.92	0.89	0.92	0.90	0.99	0.94	0.96
Females	0.89	0.96	0.92	0.92	0.89	0.90	0.94	0.99	0.96
Wght avg.	0.93	0.92	0.92	0.90	0.90	0.90	0.96	0.96	0.96
NF (PvP)	*P*	*R*	*F*	*P*	*R*	*F*	*P*	*R*	*F*
Males	0.92	0.95	0.93	0.85	0.94	0.89	1	0.91	0.95
Females	0.94	0.90	0.92	0.92	0.82	0.87	0.92	1	0.96
Wght avg.	0.93	0.93	0.93	0.88	0.88	0.88	0.96	0.96	0.96
Guk (PvE)	*P*	*R*	*F*	*P*	*R*	*F*	*P*	*R*	*F*
Males	0.95	0.93	*0.94*	0.95	0.89	0.92	0.99	0.94	0.97
Females	0.93	0.96	*0.95*	0.91	0.95	0.93	0.95	0.99	0.97
Wght avg.	0.94	0.94	*0.94*	0.93	0.93	0.93	0.97	0.97	0.97

to the class. F-measure (F) is the harmonic mean of the precision and the recall of a class, and is used as the primary indicator of a model's performance. The weighted average is the average of the prediction classes, weighted by the number of instances in each class. The confusion matrix provides an overview of how many instances were predicted accurately and inaccurately for both males and females. For each model, this is presented for the best performing, interpretable machine learning approach.

3.1 Variable Set 1 Results

1. *Overall Results:* Character gender and the number of male and female characters offer the highest information gain across all three servers for this variable set (see Table 6). The Random Tree model offers the best overall prediction results, with an f-score of 0.97 (see Table 7), but the tree leaves are not easily interpretable. As such, the best interpretable model is on the Guk Server, with a JRip f-measure of 0.94 (confusion matrix in Table 8; additional statistics in Table 9).
   a. Server Comparison: Variable Set 1 does very well across all of the different servers, with some slight differences in f-measures. Guk offers the best JRip prediction results at 0.94, one percent higher than Nagafen and two percent

Table 8 Confusion matrix Variable Set 1 Guk JRip

	Predicted male	Predicted female
Actually male	733	57
Actually female	37	811

Table 9 Model performance, Variable Set 1 Guk JRip

Number of rules: 20
Correctly classified instances: 1,544 94.26 %
Incorrectly classified instances: 9,4 5.74 %

Table 10 Prediction Variable Set 2 Variable information Gain

	Guk PvE	AB RP	NF PvP
Character gender	0.3787	0.3970	0.3200
Character race	0.0432	0.0412	0.0429
Character class	0.0296	0.0265	0.0418

higher prediction than Antonia Bayle. The lowest prediction result is the J48 on the Nagafen server, with an f-measure of 0.88. In terms of the prediction value of character gender, it makes sense that the best results are on the Guk server—PvE; it is more neutral than a PvP server, and more "identity stable" than an RP server.

b. Player Gender Comparison: The prediction results for male and female players are consistent and comparable across models and servers. Looking at the JRip results, there is no difference in the prediction results between male and female players on the Antonia Bayle server (0.92), a one percent difference in the favor of male players on the Nagafen server, and a one percent difference in the favor of female players on the Guk server. The JRip results on the Guk server, with an f-measure of 0.95 for female players and 0.94 for male players, offer the best interpretable prediction results.

3.2 *Variable Set 2 Results*

1. *Overall Results*: Character gender offers the most information gain across all three servers (Table 10). Overall, Variable Set 2 performs well across the three types of MLMs (Table 11), yielding the best results on the Antonia Bayle server with a J48 f-measure of 0.85 (confusion matrix in Table 12; additional statistics in Table 13), which is not worse than Random Tree.
   a. Server Comparison: There are no striking server differences for this variable set; Variable Set 2 performs the worst on the Nagafen PvP server with an f-measure between 0.81 and 0.82, whereas Antonia Bayle offers the highest of the three servers, with f-measures between 0.84 and 0.85.
   b. Player Gender Comparison: For female players, the J48 and JRip on the Antonia Bayle server offer the highest prediction value, with an f-measure of 0.86. For male players, an f-measure of 0.84 is the highest prediction value in this variable set, found on the Guk and Antonia Bayle servers.

Table 11 Prediction Variable Set 2

	JRip			J48			RT		
Guk (PvE)	*P*	*R*	*F*	*P*	*R*	*F*	*P*	*R*	*F*
Males	0.89	0.79	0.83	0.88	0.79	0.84	0.88	0.81	0.84
Females	0.81	0.90	0.85	0.81	0.89	0.85	0.83	0.88	0.85
Wght avg.	0.85	0.84	0.84	0.85	0.84	0.84	0.85	0.85	0.85
AB(RP)	*P*	*R*	*F*	*P*	*R*	*F*	*P*	*R*	*F*
Males	0.90	0.78	0.84	0.90	0.78	0.84	0.88	0.80	0.84
Females	0.81	0.91	0.86	0.81	0.91	0.86	0.82	0.89	0.85
Wght avg.	0.85	0.85	0.85	0.85	0.85	0.85	0.85	0.85	0.84
NF(PvP)	*P*	*R*	*F*	*P*	*R*	*F*	*P*	*R*	*F*
Males	0.81	0.83	0.82	0.81	0.83	0.82	0.82	0.80	0.81
Females	0.83	0.81	0.82	0.83	0.81	0.82	0.81	0.83	0.82
Wght avg.	0.82	0.82	0.82	0.82	0.82	0.82	0.81	0.81	0.81

Table 12 Confusion matrix Variable Set 2 AB J48

	Predicted male	Predicted female
Actually male	3,116	859
Actually female	349	3,624

Table 13 Model performance, Variable Set 2 AB J48

Number of leaves: 2
Size of the tree: 3
Correctly classified instances: 6,740 84.80 %
Incorrectly classified instances: 1,208 15.20 %

Table 14 Gender prediction Variable Set 3 variable information gain

	Guk PvE	AB RP	NF PvP
Total quests completed	0	0.00380	0.00768
Total recipes known	0	0.00568	0.02074
Total items crafted	0	0.00555	0.02927
Total rare items harvested	0	0.00247	0.01971

3.3 Variable Set 3 Results

1. *Overall Results:* The two variables that offer the most information gain (see Table 14) are total recipes known and total items crafted. Interestingly, none of these variables offer any information gain on the Guk server. Variable Set 3 does much more poorly than Variable Set 2 (see Table 15), with the highest f-measure on the Nagafen PvP server of 0.83 with Random Tree (confusion matrix in Table 16; additional statistics in Table 17). These prediction results are considerably lower than the avatar characteristics variable set.
   a. Server Comparison: Although there were no striking server differences for Variable Sets 1 or 2, for this variable set there are noticeable differences in the prediction output across servers. Variable Set 3 yields poor results on the Guk server, with an f-measure of 0.34, whereas the model predicts with an f-measure of 0.66 on the Nagafen server. This suggests that gender differences

Table 15 Prediction Variable Set 3

	JRip			J48			RT		
Guk (PvE)	*P*	*R*	*F*	*P*	*R*	*F*	*P*	*R*	*F*
Males	0.55	0.59	0.57	0.50	1	0.67	0.83	0.73	0.78
Females	0.55	0.50	0.52	0	0	0	0.76	0.85	0.80
Wght avg.	0.55	0.55	0.55	0.25	0.50	0.34	0.80	0.79	0.79
AB(RP)	*P*	*R*	*F*	*P*	*R*	*F*	*P*	*R*	*F*
Males	0.56	0.57	0.56	0.54	0.71	0.61	0.79	0.75	0.77
Females	0.56	0.55	0.55	0.58	0.40	0.48	0.76	0.81	0.78
Wght avg.	0.56	0.56	0.56	0.56	0.56	0.55	0.78	0.78	0.78
NF(PvP)	*P*	*R*	*F*	*P*	*R*	*F*	*P*	*R*	*F*
Males	0.59	0.57	0.58	0.66	0.65	0.65	0.83	0.83	0.83
Females	0.59	0.61	0.60	0.66	0.67	0.66	0.83	0.83	0.83
Wght avg.	0.59	0.59	0.59	0.66	0.66	0.66	0.83	0.83	0.83

Table 16 Confusion matrix Variable Set 3 NF RT

	Predicted male	Predicted female
Actually male	1,396	766
Actually female	727	1,469

Table 17 Model performance, Variable Set 3 NF RT

Number of leaves: 127
Size of the tree: 253
Correctly classified instances: 2,865 65.74 %
Incorrectly classified instances: 1,493 34.26 %

for these behavioral indicators are less pervasive on the PvE server, and much more so on the PvP server.

b. Player Gender Comparison: Variable Set 3 shows notable prediction output differences between male and female players. Most noticeable are the results on the Guk server, which shows no prediction output for females with the J48 and an f-measure of 0.67 for male players. Also, on the Antonia Bayle server, the J48 offers an f-measure of 0.61 for male players and 0.48 for female players. The other results are more similar between males and females. The best interpretable prediction results for male players are on the Guk server with an f-measure of 0.67, and for female players it is on the Nagafen server with an f-measure of 0.66.

3.4 Variable Set 4 Results

1. *Overall Results:* The information gain for each variable is the same as it was in the previous variable set in which it was included. Table 18 shows the combined information gain for this variable set. Avatar characteristics offer higher/more information gain than gameplay behaviors do. Results for Variable Set 4 (see Table 19) show that this variable set does slightly better than Variable Set 2, and

Table 18 Prediction variable Set 4 Variable information Gain

	Guk PvE	AB RP	NF PvP
Character gender	*0.3787*	*0.3970*	*0.3200*
Character race	0.0432	0.0412	0.0429
Character class	0.0296	0.0265	0.0418
Total items crafted	0	0.00555	0.02927
Total recipes known	0	0.00568	0.02074
Total rare items harvested	0	0.00247	0.01971
Total quests completed	0	0.00380	0.00768

Table 19 Prediction Variable Set 4

	JRip			J48			RT		
Guk (PvE)	*P*	*R*	*F*	*P*	*R*	*F*	*P*	*R*	*F*
Males	0.89	0.79	0.84	0.88	0.79	0.84	0.95	0.89	0.92
Females	0.81	0.90	0.85	0.81	0.89	0.85	0.89	0.95	0.92
Wght avg.	0.85	0.84	0.84	0.85	0.84	0.84	0.92	0.92	0.92
AB(RP)	*P*	*R*	*F*	*P*	*R*	*F*	*P*	*R*	*F*
Males	0.90	0.79	0.84	0.90	0.78	0.84	0.94	0.88	0.91
Females	0.81	0.91	0.86	0.81	0.91	0.86	0.88	0.94	0.91
Wght Avg.	0.86	0.85	0.85	0.85	0.85	0.85	0.91	0.91	0.91
NF(PvP)	*P*	*R*	*F*	*P*	*R*	*F*	*P*	*R*	*F*
Males	0.86	0.79	0.83	0.92	0.80	0.85	0.98	0.89	0.93
Females	0.81	0.87	0.84	0.82	0.93	0.87	0.90	0.98	0.94
Wght Avg.	0.84	0.83	0.83	0.87	0.86	0.86	0.94	0.93	0.93

much better than Variable Set 3. Variable Set 4 performs best on the Nagafen sever, with an f-measure of 0.86 using J48 (confusion matrix in Table 20; additional statistics in Table 21) and 0.93 using Random Tree.

a. Server Comparison: Similar to Variable Set 2, there are no striking server differences. Variable Set 4 performs best on the Nagafen PvP server, with the Guk PvE server yielding slightly lower results. The PvP server appears to offer the highest prediction results for these variables, again suggesting that male and female players engage in gendered behaviors on this server type.
b. Player Gender Comparison: Unlike Variable Set 3, Variable Set 4 shows only slight differences in the prediction output for male and female players. Unlike Variable Set 2, Variable Set 4, on the Nagafen PvP server, offers the highest f-measures for both male and female players, 0.87 for female players and 0.85 for male players with J48. The Random Tree output for this variable set offers an f-measure of 0.94 for female players and 0.93 for male players. However, this tree is much more challenging to interpret than the previous Random Tree model because of the numerous detailed leaves contained in Variable Set 3's output.

Table 20 Confusion matrix Variable Set 4 NF J48

	Predicted male	Predicted female
Actually male	1,722	440
Actually female	151	2,045

Table 21 Model performance, Variable Set 4 NF J48

Number of leaves: 394
Size of the tree: 519
Correctly classified instances: 3,767 86.44 %
Incorrectly classified instances: 591 13.56 %

3.5 Recap and Overview of Variable Set Results

Variable Set 1 with combined avatar characteristics and gameplay for primary characters provides the best interpretable prediction results at 0.94. The Variable Set 1 results show that for both male and female players, character gender, and the numbers of male and female characters are the most significant gameplay behavioral indicators in EQII to predict player's gender. This variable set performs better than Variable Sets 2–4, suggesting that players' decisions about and behaviors with their primary characters is more reliable and indicative of stable elements of their identity (e.g., gender) than such decisions and behaviors with nonprimary characters.

Out of Variable Sets 2, 3 and 4, which include avatar characteristics and/or gameplay behaviors for all characters played, Variable Set 4, with both variable types, offers the best interpretable results at 0.83 with JRip on the Nagafen (PvP server. This suggests that gender differences are more pronounced on PvP servers than PvE or RP servers. Moreover, the results for Variable Sets 2–4 show the importance of avatar characteristics in the prediction of player gender; the avatar variable set does much better than the gameplay variable set in predicting player gender, with character gender as the strongest predictor.

3.6 Prediction Lay-statements

J48 models provide pruned trees for interpretation, similar to JRip rulesets with leaves acting as if/then statements. Random Tree algorithms offer very detailed and complex trees that are challenging to interpret. This section provides an overview of what the rulesets and tree leaves offer, in addition to some exemplar lay statements that span all variable sets.

1. *Variable Set 1 Overview:* The rulesets in Variable Set 1 show the high prediction value of primary character gender, and the number of male and female characters a player has. Although there are other variables in the variable set, these offer the most prediction value. Typically, players choose matched gender characters, and this is also reflected in the number of male and female characters a player creates. The rulesets with the most coverage do not offer too much information

Table 22 Character gender as player gender predictor in JRip rulesets across servers (Variable Set 2)

	Rule/Leaf	NF PvP	Guk PvE	AB RP
Male character	Accuracy	81 % M	88 % M	90 % M
	Coverage	60 %	50 %	48 %
	Applies to	$n = 2{,}211$	$n = 2{,}738$	$n = 3{,}465$
Female character	Accuracy	83 % F	81 % F	81 % F
	Coverage	58 %	65 %	67 %
	Applies to	$n = 2{,}147$	$n = 3{,}338$	$n = 4{,}483$

about play styles; quests and deaths are noted in a few rulesets, but there are no interpretable patterns.

2. *Variable Set 1 Statements—Predicting Males:*
   - If a character on a PvE server is male, the player has one or less female characters, and 246 or more quests completed, 99 % will be men in real life ($n = 352$, 352/5, Accuracy = 99 %, Coverage = 21 %). JRip 0.94
   - If a character on a PvE server is male, and the players has one or less female characters, 96 % will be men in real life ($n = 541$, 541/19, Accuracy = 96 %, Coverage = 33 %). J48 0.93
   - If a character on an RP server is male, and they have more than 44 total quests, 96 % will be men in real life ($n = 906$, 906/37, Accuracy = 96 %, Coverage = 42 %). J48 0.90
   - If a character on a PvP server is male, the player has no female characters, and has 325 or less total deaths, 97 % will be men in real life ($n = 105$, 105/3, Accuracy = 97 %, Coverage = 22 %). JRip 0.93
   - If a character on a PvP server is male and the player has no female characters, 89 % will be men in real life ($n = 151$, 151/16), Accuracy = 89 %, Coverage = 31 %). J48 0.88
3. Variable Set 1 Statements—Predicting Females:
   - If a character on a PvE server is female, the player has more than three female characters, and no male characters, 96 % will be women in real life ($n = 528$, 528/19, Accuracy = 96 %, Coverage = 32 %). J48 0.93
   - If a character on an RP server is female, the player has one or less male characters, and 80 or more total deaths, 91 % will be women in real life ($n = 638$, 638/55, Accuracy = 91 %, Coverage = 30 %). JRip 0.92
   - If a character on a RP server is female and the player has no male characters, 92 % will be women in real life (n=778, 778/64, Accuracy = 92 %, Coverage = 36 %). J48 0.90
   - If a character on a PvP server is female, 85 % will be women in real life ($n = 232$, 232/35, Accuracy = 85 %, Coverage = 48 %). J48 0.88
4. *Variable Sets 2–4 Overview:* The rulesets, pruned trees, and leaves in Variable Sets 2–4 show some of the same patterns as in Variable Set 1. In particular, character gender offers the most predictive value for player gender, which is consistent across all three servers. Table 22 shows an overview of JRip rulesets across three servers, and the accuracy and coverage of character gender in Variable Set 2.

Although Variable Sets 2–4 do not represent our best prediction results, these variable sets do offer insight into gameplay trends that Variable Set 1 did not offer. Female players are more consistent in their immersive gameplay, as they craft, scribe recipes, and harvest items across all three servers in varying amounts. This differs from male players, who do not engage in the same immersive play behaviors. For example, male players show low immersive gameplay statistics on the Nagafen PvP server, where male players are likely engaging in combat activities.

5. *Variable Sets 2–4 Statements—Predicting Males from Character Gender:*
   - If a character on a PvP server is male, then it is 81 % likely the player will be male in real life ($n = 2{,}211$, 2,211/417, Accuracy = 81 %, Coverage = 60 %). JRip 0.82 and J48 0.82
   - If a character on a PvE server is male, then it is 88 % likely the player will be male in real life ($n = 2{,}738$, 2,738/324, Accuracy = 88 %, Coverage = 50 %). J48 0.84
   - If a character on a RP server is male, then it is 90 % likely the player will be male in real life ($n = 3{,}465$, 3,465/349, Accuracy = 90 %, Coverage = 48 %). J48 0.85
6. *Variable Sets 2–4 Statements—Predicting Males from low crafting, scribing, and questing:*
   - If a player on a PvP server has 16 or less items crafted, then it is 58 % likely the player is male in real life ($n = 2{,}120$, 2,120/889, Accuracy = 58 %, Coverage = 69 %). JRip 0.59
   - If a character on a PvP server is male, and they do not know any recipes, then it is 78 % likely the player is male in real life ($n = 622$, 622/136, Accuracy = 78 %, Coverage = 17 %). JRip 0.83
   - If a player on a PvP server player has crafted five or less items, knows no recipes, has completed 17 or less quests, then it is 61 % likely the player is male in real life ($n = 776$, 776/302, Accuracy = 61 %, Coverage = 25 %). J48 0.66
7. *Variable Sets 2–4 Statements—Predicting Males from high crafting and questing on an RP server:*
   - If a player on an RP server has crafted 196 items or less, has not harvested any rare items, but has questing activity, then it is 55 % likely the player is male in real life ($n = 2{,}440$, 2,440/1,091, Accuracy = 55 %, Coverage = 42 %). J48 0.55
8. *Variable Sets 2–4 Statements—Predicting Females from Character Gender:*
   - If a character on a PvP server is female, then it is 83 % likely the player will be female in real life ($n = 2{,}147$, 2,147/368, Accuracy = 83 %, Coverage = 58 %). J48 0.82
   - If a character on a PvE server is female, then it is 81 % likely the player will be female in real life ($n = 3{,}338$, 3,338/633, Accuracy = 81 %, Coverage = 65 %). JRip 0.84 and J48 0.84
   - If a character on an RP server is female, then it is 81 % likely the player will be female in real life ($n = 4{,}483$, 4,483/859, Accuracy = 81 %, Coverage = 67 %). JRip 0.85 and J48 0.85

9. *Variable Sets 2–4 Statements—Predicting Females from Character Gender and High Recipe Scribing on a PvP Server:*
   - If a character on a PvP server is female, and she knows 462 or more recipes, then it is 89 % likely the player if female in real life ($n = 1{,}024$, 1,024/116, Accuracy = 89 %, Coverage = 26 %). J48 0.86
10. *Variable Sets 2–4 Statements—Predicting Females from Crafting, Questing, and Recipe Scribing on an RP Server:*
    - If a player on a RP server has one or more rare items harvested, has crafted 33 or less items, and has completed 120 or less quests, then it is 56 % likely the player is female in real life ($n = 864$, 864/376, Accuracy = 56 %, Coverage = 16 %). JRip 0.56

4 Discussion

The present research examined the prediction of player gender in the game EQII using data provided by the game operator from over 4,000 players. Machine learning models provided high f-measures (up to 0.97) as well as interpretable statements that could be used to understand the patterns in the data, generally with the Random Tree models contributing more to the former and JRip and J48 more to the latter, as expected. Avatar characteristics, compared to gameplay behaviors, provided the highest predictive value, with players generally choosing primary characters as well as multiple nonprimary characters to match their gender, though crafting, scribing, harvesting, and questing also emerged as significant variables in some of the models. Further, predictions based on primary characters were stronger than those that also considered nonprimary characters, suggesting that uses of primary characters are more stable and reflective of offline identity traits (e.g., gender), while uses of nonprimary characters are more exploratory and thus less predictive of offline identity traits. Most interestingly, strength of the predictive models varied in notable ways across the RP, PvP, and PvE servers that were examined.

These differences between servers in the predictive models likely occurred because of differences in rule structures of each server. PvE is the most general style of MMO play, where the focus is on exploring, questing, or raiding in groups, but where players are not able to attack one another (i.e., they can only attack elements of the environment). PvP servers allow the players to combat each other, and while the game still includes noncombat activities such as questing and scribing, the potential for combat reduces the focus on these other activities. RP servers are like PvE servers, except there is an understanding that players may roleplay their characters, i.e., pretend to be the character by communicating and acting in a fashion that befits the character's identity in the fictional game space, though in practice the proportion of people who treat such roleplaying seriously is quite small [10]. Given the large differences in these frameworks of "play," it makes sense that different server types would attract different types of people and encourage different types of gendered play behavior.

The consideration of such differences provides some insight into the variance in strength of the predictive models. For one, the variable set with only avatar variables performed best on the RP server, suggesting that players are more likely to choose avatars that match their offline gender on RP servers. While this may seem counter-intuitive initially (i.e., why roleplay a character similar to yourself), it makes sense given that people are generally poor at maintaining a false identity that is far from their own [11], and so players would have an easier time roleplaying a character that matches their gender.

Similarly, the gameplay variables offer more prediction value on the PvP server than the PvE server. Specifically, on the PvP server, a player with low crafting, scribing, and questing was likely to be male, suggesting that males on PvP servers spend more time engaging in combat. Interesting, a high amount of recipe scribing helped predict female players on the PvP server, suggesting that females do not focus on combat in such spaces, but perhaps support the players who do by contributing to the in-game economy. This reasoning is also supported by the fact that in Variable Set 4, female players are easiest to predict on the PvP server.

5 Conclusion

Overall, these findings support previous research about gendered gameplay in MMOs. Male and female players' character choices tend to match their gender, and their play styles tend to be consistent with gendered stereotypes. Namely, female players craft, harvest, quest, and scribe recipes more than male players do; male players engage in combat and killing more than female players do. Unlike virtual havens where people "leave the meat behind" and become anyone or anything, MMO players appear to use the game as a space to embody their offline identities. Of course, the analyses presented here are based on data from one MMO, and so generalizations should be treated with caution and future research should aim to corroborate and extend these findings. Still, the present research contributes a new methodological approach (machine learning models), framework for variable analysis (avatar characteristics vs gameplay behaviors), and contextual understanding (server-based rulesets) to this area of research.

Acknowledgments The research reported herein was supported by the National Science Foundation (NSF) via award number: IIS-0729421, the Army Research Institute (ARI) via award number W91WAW-08-C-0106, Air Force Research Lab (AFRL) via Contract No: EA8650-10-C-7010 and the Army Research Lab (ARL) Network Science–Collaborative Technology Alliance (NSCTA) via BBN TECH/W911NF-09-2-0053. The data used for this research were provided by the SONY Online Entertainment (SONY Corporation). We gratefully acknowledge all our sponsors. The findings presented do not in any way represent, either directly or through implication, the policies of these organizations.

References

1. Huh, S., Williams, D.: Dude looks like a lady: Gender swapping in an online game. In: Bainbridge, W. (ed.) Online Worlds: Convergence of the Real and the Virtual. pp. 161–174, Springer, London (2010)
2. Yee, N., Ducheneaut, N., Yao, M., Nelson, L.: Do men heal more when in drag?: Conflicting identity cues between user and avatar. Proceedings of the SIGCHI Conference on Human Factors in Computing Systems. ACM (2011)
3. Hussain, Z., Griffiths, M.D.: Gender swapping and socializing in cyberspace: An exploratory study. CyberPsychol. Behav. **11**(1):47–53 (2008)
4. Roberts, L.D., Parks, M.R.: The social geography of gender-switching in virtual environments on the Internet. Inf. Commun. Soc. **2**(4):521–540 (1999)
5. MacCallum-Stewart, E.: Real boys carry girly epics: Normalising gender bending in online games. Eludamos J. Comput. Game Cult. **2**(1):27–40 (2008)
6. Yee, N.: Maps of digital desires: Exploring the topography of gender and play in online games. In: Kafai, Y.B., Heeter, C., Denner, J., Sun, J.Y. (eds.) Beyond Barbie and Mortal Kombat: New Perspectives on Gender and Gaming. pp. 83–96, MIT Pres, Cambridge (2008)
7. Hall, M., Frank, E., Holmes, G., Pfahringer, B., Reutemann, P., Witten, I.H.: The WEKA data mining software: An update. SIGKDD Explor. **11**(1) (2009)
8. Pang-Ning, T., Steinbach, M., Kumar, V.: Introduction to Data Mining. Addison-Wesley Longman Publishing, Boston (2005)
9. Russell, S., Norvig, P.: Artificial Intelligence: A Modern Approach. Prentice Hall, Upper Saddle River (2010)
10. Williams, D., Kennedy, T.L.M., Moore, R.J.: Behind the avatar: The patterns, practices, and functions of role playing in MMOs. Games Cult. **6**(2):171–200 (2011)
11. Donath, J.S.: Identity and deception in the virtual community. In: Smith, M.A., Kollock, P. (eds.) Communities in Cyberspace, pp. 29–59, Routledge, London (1999)

Predicting Links in Human Contact Networks Using Online Social Proximity

Annalisa Socievole, Floriano De Rango and Salvatore Marano

Abstract Experimentally measured contact traces, such as those obtained through short range wireless sensors, have allowed researchers to study how mobile users contact each other in different environments. These traces often include other types of useful information such as users' social profiles and their online friend lists. This explicit social information is important since it can be exploited for augmenting the knowledge of user behavior and hence improve the quality of human mobility analysis. In this paper, we use online social ties for predicting users' contacts. Specifically, we study the prediction of links in human contact networks as a graph inference problem, where the existence of an edge is predicted using different proximity measures that quantify the closeness or similarity between nodes. First, we predict the edges of the contact graph when we have only information about users' online social network. Next, we analyze the effectiveness of using both the online social network and a part of the contact network for contact prediction. In both settings, our study on three different human contact traces shows that resource allocation measure plays a significant role in contact prediction. Furthermore, the results demonstrate the importance of online social proximity in identifying stronger ties.

1 Introduction

The recent growth of ubiquitous systems, mobile social media applications, and social data has led to an increasing interest in the analysis of social networks. Understanding the dynamics of our social networks is a challenging task considering that

A. Socievole(✉)
Dipartimento di Ingegneria Informatica,
Modellistica, Elettronica e Sistemistica (DIMES), Università della Calabria,
Ponte P. Bucci, 87036 Rende, Italia
e-mail: socievolea@dimes.unical.it

F. De Rango
Dipartimento di Ingegneria Informatica, Modellistica, Elettronica e Sistemistica (DIMES),
Università della Calabria, Ponte P. Bucci, 87036 Rende, Italia
e-mail: derango@dimes.unical.it

S. Marano
Dipartimento di Ingegneria Informatica, Modellistica, Elettronica e Sistemistica (DIMES),
Università della Calabria, Ponte P. Bucci, 87036 Rende, Italia
e-mail: marano@dimes.unical.it

M. A. Ahmad et al. (eds.), *Predicting Real World Behaviors from Virtual World Data,*
Springer Proceedings in Complexity, DOI 10.1007/978-3-319-07142-8_6,

these networks are complex and consist of several overlapping parts. For example, people may be connected both on online social networks (e.g., Facebook, Twitter, LinkedIn, and YouTube) and in the real world through different types of relationships such as friendships, family, and colleagues. In contrast to online networks, human contact networks are usually acquired through short-range wireless sensors or Bluetooth enabled devices that record when mobile users are in close proximity. The resulting contact traces containing colocation data has allowed researchers to investigate human mobility and design more realistic mobility models (e.g., [7, 16]). This in turn has influenced routing algorithms proposed for delay-tolerant networks (DTNs) [12] where human contacts play a fundamental role in message delivery. In such networks, also known as pocket-switched networks (PSNs) [8], nodes can exchange messages when a contact opportunity arises. Hence, successful delivery strongly relies on human contact patterns.

While there are many studies that analyze separately online and offline social networks of human contacts, some important aspects dealing with the social network which combines both online and offline connections for a given set of users still remain unexplored. The analysis of such a network can provide more information concerning user behavior and consequently on how links get established. In this paper, we are interested in predicting mobility in human contact networks exploiting both online and offline social information. To the best of our knowledge, our work is the first one which addresses the problem of predicting the formation of links in a human contact network by using the corresponding online social network. In the literature, this problem is often addressed by using measures which compute the similarity of nodes in the graph representing the contact network. According to the homophily theory [15], similar people are indeed more likely to interact with each others. Thus, the evolution of the contact network can be inferred using features that are inherent to the structural properties of the contact graph.

In this work, we consider three real-world datasets of human mobility containing also the Facebook friend lists of the mobile users. In the first part of the paper, for each dataset, we start analyzing the similarities between the contact network graph constructed from mobility data and the corresponding online social network graph constructed from Facebook friend lists. Then we consider a set of proximity measures and compute them both on the contact network graph and on the online social network graph in order to understand if these measures well correlate with contact duration and the number of contacts. In the second part of the paper, we formulate the problem of link prediction in human contact networks as a graph inference problem. First, we assume that we have only information about the links in the online social network graph while those in the contact network graph are completely unknown. Then, we study a similar problem in a different setting where we try to infer the contact graph using both the online social network and part of the contact network.

Our contributions can be summarized as follows:

- We consider three human contact traces related to different experiments and show the relation between people's contact patterns and their online social network.
- We present two methods for predicting links in human contact networks when we have only knowledge of people's online social network and when both online social network and a part of contact network are known.

- We demonstrate that *resource allocation* is the most significant feature in contact prediction.
- Finally, we show that node similarity computed on a graph that fuses part of contact graph with online social graph better predicts stronger ties.

The remainder of the paper is structured as follows. Section 2 reviews the state of the art in this research area. Section 3 discusses the motivation of our work. Section 4 describes the problem to be tackled. Section 5 presents the performance results of our method of contact graph inference using online social information. Section 6 shows the results of the contact graph inference using partial contact information. Finally, Sect. 7 concludes the paper.

2 Related Work

In this section, we discuss the related work concerning the comparison between the network of physical encounters and the online social network for a given set of users, and the link prediction in the context of human contact networks.

2.1 Analysis of Contact Graph and Corresponding Online Social Graph

Bigwood et al. [3] explored the use of the social network Facebook instead of the detected social network discovered through encounters for routing in ubiquitous computing environments, showing that the online social information decreases the delivery cost and produces comparable delivery ratio. They also analyzed the two networks in terms of structural equivalence and role equivalence and discovered that they differ. Socievole and Marano [22] compared the same contact and online networks in terms of egocentric and sociocentric behaviors. Analyzing node centrality and communities, they showed that the two networks differ except for betweenness centrality that is similar in both networks. In another work [6], Ciobanu et al. found that the contact and Facebook social graphs in an academic environment have a high degree of structural similarity. In [19], Pietilänen and Diot showed that the temporal communities of contact network exhibit a high level of correlation with social characteristics such as online friendship, shared affiliations, home city, or country of origin.

Kostakos and Venkatanathan [13] have tackled the fusion of contact and online networks highlighting the high-level structural similarities between the two types of networks, and noting the underlying differences in how individuals take part in these social networks. In a similar work [18], Pan et al. showed that the contact and online social networks represent two different classes of social engagement that complement each other.

2.2 Link Prediction in Human Contact Networks

Song et al. [23] studied the limits of predictability in a network of mobile phone users, finding a 93 % potential predictability in user mobility due to its high degree of regularity. Using mobile phone data as well, Wang et al. [25] found a strong correlation between users' movements and their connections in the social network. In [24], Vu et al. developed a model to predict mobile users' future locations and encounters by exploiting the regularity of mobility found in a real joint WiFi/Bluetooth trace. Scholz et al. [21] analyzed the problem of predicting new links and recurring links in a human contact network acquired through a radio frequency identification (RFID)-based system recording face-to-face interactions. Their analysis' results provide interesting insights concerning the impact of the contact durations and the strength of such stronger ties.

Jahanbakhsh et al. [9] formulated the problem of human contact prediction as a graph inference problem. They showed the importance of offline social profiles (nationality, spoken languages, current affiliations, city and country of residence, school, and research interests) and partial knowledge of the contact graph as well for link prediction. The same authors evaluated the performance of several methods for predicting the missing contacts in different social environments, showing that the combination of the number of common neighbor feature with social data provides the best predictions [10]. Successively, they employed a supervised learning approach for contact prediction, showing that the number of common neighbor and total overlap time play essential roles in forming human contacts [11]. Moreover, they showed that nodes with high centralities are more predictable.

3 Preliminary Observations

As a first step, we discuss the motivation of our work. Considering the importance of the homophily theory in the link formation process in social networks, we want to test the power of different social similarity measures for predicting the contacts among nodes in a mobile network. According to this theory, in fact, individuals who are socially similar are more likely to contact each other than those who are not similar. Hence, we start analyzing the influence of online and offline social similarities between nodes on their total contact durations and their total number of contacts. First, we define the node-pairs features used to measure the social similarity. Next, we describe the datasets used and finally, we show the results of the analysis.

3.1 Social Similarity Measures

As the main focus of the paper is to explore the predictive power of similarity between nodes, we selected five representative measures which have been proven to perform reasonably well in previous studies [14, 21, 25].

- **Common Neighbors** This measure is based on the assumption that it is more likely that two nodes are connected if they have many neighbors in common. For two nodes i and j having a set of neighbors N_i and N_j, respectively, this measure is defined as

$$sim_{CN}(i,j) = \left|N_i \cap N_j\right| \tag{1}$$

- **Jaccard Coefficient** It is a normalized version of the common neighbors measure. Two nodes may have many common neighbors because each one has a lot of neighbors, not because they are strongly related to each other. It is defined as

$$sim_{JC}(i,j) = \frac{\left|N_i \cap N_j\right|}{\left|N_i \cup N_j\right|} \tag{2}$$

- **Adamic Adar** [1] Similar to common neighbors, this measure gives more weight to neighbors that are not shared with many others

$$sim_{AA}(i,j) = \sum_{z \in N_i \cap N_j} \frac{1}{log\,|N_z|} \tag{3}$$

- **Resource Allocation** [26] This measure is motivated by the resource allocation dynamics on complex networks. If we consider a pair of nodes, i and j, which are not directly connected, i can send some resource to j through their common neighbors. The similarity between i and j can be defined as the amount of resource that j received from i, which is

$$sim_{RA}(i,j) = \sum_{z \in N_i \cap N_j} \frac{1}{|N_z|} \tag{4}$$

- **Preferential Attachment** [2] The mechanism of preferential attachment can be used to generate evolving scale-free networks, where the probability that a node is connected to node i is proportional to the degree of i. Hence, the similarity between i and j is defined as

$$sim_{PA}(i,j) = \left|N_i\right|\left|N_j\right| \tag{5}$$

3.2 *Datasets*

In this paper, we use three experimental datasets of human mobility that are publicly available: UPB [5], Sassy [4], and Sigcomm [20]. Table 1 summarizes their main characteristics.

UPB is a dataset related to an Android Bluetooth tracing experiment performed for a period of 35 days in an academic environment (University Politehnica of Bucharest).

Table 1 Summary of the datasets used for link prediction

	UPB	Sassy	Sigcomm
# of devices	22	25	76
Device type	Phone	T-mote	Phone
Duration	35 days	79 days	3 days
Environment	Academic	Academic	Conference
Contact type	Bluetooth	ZigBee	Bluetooth
Radio range	10 m	10 m	10–20 m
Granularity	60–3600 s	6.67 s	120 ± 10.24 s

Data were collected using an Android application[1] that logged Bluetooth contacts within a range of 10 m between 22 mobile phones. The polling interval (between 5 and 30 min) for each device could be controlled by the user.

Sassy is a dataset of encounter records and corresponding Facebook social network data of a group of participants at University of St. Andrews. Twenty-five individuals carried IEEE 802.15.4 T-mote sensor nodes for 79 days in order to collect colocation data. The ZigBee devices were able to detect each other within a radius of 10 m and were programmed to broadcast a beacon every 6.67 s.

Sigcomm dataset includes Bluetooth copresence data collected by the opportunistic mobile social application MobiClique and the social profiles (Facebook friends and interests) of the participants during the SIGCOMM 2009 conference. Seventy-six HTC s620 Windows Mobile smartphones were distributed to a set of volunteers during the first days of the conference, performing a periodic Bluetooth device discovery every 120 ± 10.24 s. The experimental devices had a class 2 Bluetooth v2.1 radio with a range of 10–20 m.

In Figs. 1 and 2, the contact duration and the number of contacts distributions are presented, respectively. As can be seen, they follow both an approximate power law in each dataset. Figure 1 represents the distribution of contact durations between node pairs. In academic scenarios (UPB and Sassy), the distributions are quite similar. Only 25 % of contact durations last more than 2 min. On the contrary, in the conference scenario of Sigcomm where participants are colocated attending sessions most of the time, contact durations are longer. Here, 50 % of contact durations last more than 2 min. Looking at Fig. 2 which represents the distribution of the total number of contacts between each possible node pair, Sassy shows a number of contact opportunities greater than Sigcomm and UPB since its trace duration is longer (79 days). On the contrary, despite the fact that UPB trace duration is significantly longer than Sigcomm (35 vs. 3 days), UPB distribution shows a low number of contacts since contact opportunities are more sporadic.

3.3 *Impact of Social Similarity on Human Contacts*

Before analyzing the influence of online and offline social similarities on the contacts between nodes, we study the structural commonality between social information

[1] http://code.google.com/p/social-tracer/.

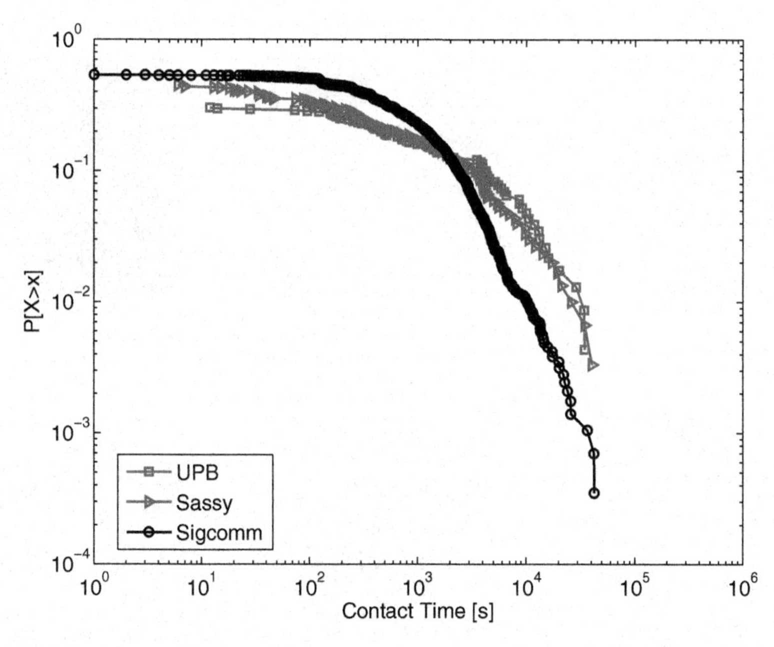

Fig. 1 Distribution (complementary cumulative distribution function, CDF) of contact durations

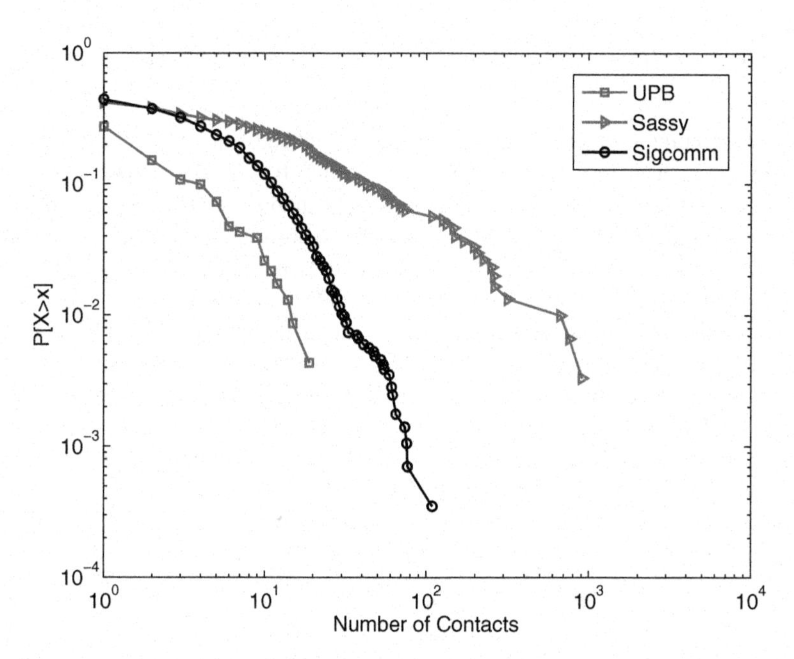

Fig. 2 Distribution (complementary cumulative distribution function, CDF) of the number of contacts

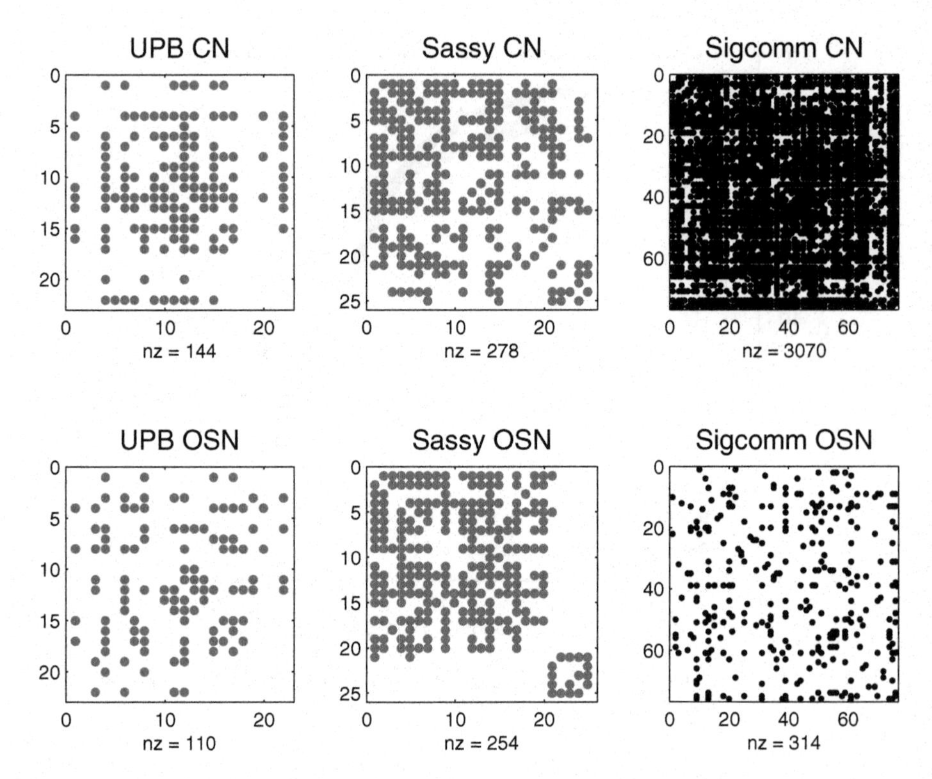

Fig. 3 Spy plots of contact network (*CN*) and online social network (*OSN*) adjacency matrices

defined by online social networks and contacts created by human mobility. Since the datasets do not have the accurate social relationships between participants, we use the contact data to generate a contact network graph. In this graph, an edge between two nodes is present if these nodes meet at least once during the trace period. Similarly, we use the participants' Facebook friend lists to generate an online social network graph, where an edge between two nodes exists if they are friends. From these graphs, we form two adjacency matrices for each dataset and depict them in the spy plots of Fig. 3, where each nonzero entry in the matrices corresponds to a point in the plot. As can be observed, the number of offline contacts is greater than online friendships since it deals with mobility data. Moreover, while UPB and Sassy show a comparable number of offline and online contacts, the Sigcomm contact network matrix is much more dense than the online social network matrix (3070 vs. 314 nonzero entries).

In order to quantify how closely the contact graph and the online social graph are matching, we measure the *closeness error* (CE) [17] between them. Given two graphs $G(V, E)$ and $G'(V', E')$, CE is defined as follows:

$$CE(G, G') = \frac{\left|E \setminus E'\right| + \left|E' \setminus E\right|}{|E \cup E'|} \qquad (6)$$

Table 2 Closeness error between contact network and online social network graphs

	UPB	Sassy	Sigcomm
$CE(G_{CN}, G_{OSN})$	0.6489	0.5851	0.9278

Table 3 Correlation between different predictor scores and contacts duration (*CD*), and number of contacts (*NC*). Common neighbors (*CN*), Jaccard coefficient (*JC*), Adamic Adar (*AA*), resource allocation (*RA*) and preferential attachment (*PA*) are computed on contact network graph

	UPB		Sassy		Sigcomm	
	CD	NC	CD	NC	CD	NC
sim_{CN}	0.4092	0.5857	0.2973	0.1565	0.1340	0.4718
sim_{JC}	0.3073	0.48	0.2511	0.1276	0.1556	0.4573
sim_{AA}	0.4564	0.6192	0.3152	0.1673	0.1363	0.4783
sim_{RA}	0.5035	0.6506	0.3388	0.1818	0.1436	0.4968
sim_{PA}	0.4623	0.6204	0.3055	0.1581	0.1233	0.4698

Table 4 Correlation between different predictor scores computed on online social network and levels of interactions on contact network

	UPB		Sassy		Sigcomm	
	CD	NC	CD	NC	CD	NC
sim_{CN}	0.3796	0.4074	−0.0167	−0.0111	0.0479	0.1238
sim_{JC}	0.2316	0.2616	0.0334	0.0422	0.0656	0.1445
sim_{AA}	0.4469	0.4665	0.0372	0.0601	0.0577	0.1376
sim_{RA}	0.4840	0.5033	0.1096	0.1511	0.0641	0.1448
sim_{PA}	0.3555	0.3890	−0.0396	−0.0415	0.0249	0.0888

Dealing with the fraction of dissimilar edges in the two graphs, CE takes values between 0 and 1. The more the graphs differ, the closer the CE value is to 1. Table 2 shows the CE values for the different datasets. Sassy gives the best matching with the contact graph. This means that the online friendship is strongly correlated with the offline encounters.

We now focus on comparing the influence of offline and online social similarity on contact durations and on the number of contacts. Table 3 shows the Pearson correlation coefficient between the social similarity of all pairs of nodes measured on contact network graph and their total number of contacts, and total durations. The positive dependency obtained from the correlations between the contact pattern for a pair of nodes and their social similarity demonstrates the potential of the considered similarity measures as predictors of physical meetings. *Resource allocation* shows the highest Pearson coefficients across all datasets except for Sigcomm when contact duration is considered. In this case, the *Jaccard coefficient* performs best. Note that *Adamic Adar* and *preferential attachment* are also good predictors.

In Tables 3 and 4, we present the results of the correlation between similarity computed on the online social network graph and the levels of interactions between nodes. In this case, the correlation values between similarity and the number of contacts are the highest. As in the previous case, *resource allocation* performs best across all datasets except for Sigcomm that for contact duration gives again the

Table 5 Euclidean distances between contact network and online social network similarities

	UPB	Sassy	Sigcomm
$dist_{CN}$	51.97	81.3941	1.5255×10^3
$dist_{JC}$	4.5198	4.5455	23.9473
$dist_{AA}$	22.5714	31.9756	391.0915
$dist_{RA}$	5.3869	6.5864	31.6165
$dist_{PA}$	703.2354	1.5486×10^3	1.0104×10^5

best results for the *Jaccard coefficient*. For UPB dataset, both *Adamic Adar* and *common neighbors* perform well, while for Sassy and Sigcomm both *Adamic Adar* and *Jaccard Coefficient* are good predictors.

In Table 5, the Euclidean distances between the values of each similarity measure computed on the contact network and online social network graphs are shown. We can observe that UPB shows the lowest distances across all measures. This result confirms the highest correlations between each similarity measure computed on the online social network graph and the levels of interactions between nodes. We can further observe that *Jaccard coefficient* and *resource allocation* show low distance values across all datasets, while there is a significant difference between *preferential attachment* similarity values computed offline and online. Considering that this last measure is based on nodes' degrees, this result can be explained looking at Fig. 4, where the correlation between contact network and online social degrees is shown. Degree varies among nodes: while the correlation between the two measures in UPB and Sassy is medium (0.5205 and 0.4177, respectively), in Sigcomm is low (0.1605).

4 Problem Definition

In this section, we present the problem formulation. Let $G_{CN} = (V, E_{CN})$ denote an undirected contact network graph where an edge $e_{CN} = (u, v)$ between two nodes u and v exists if they had at least one contact during $[t_s, t_e]$, where t_s is the starting time of the experimental trace and t_e is its end time. Let $G_{OSN} = (V, E_{OSN})$ denote the undirected online social network graph where an edge $e_{OSN} = (u, v)$ indicates an online friendship between u and v. We consider two prediction problems. First, we assume that the edges in $E_{CN} \subset V \times V$ are totally unknown. Thus, the first problem is to infer edges in E_{CN} using available information about edges in E_{OSN}.

The second prediction problem is to predict the contact graph using both the online social network and part of the contact network. Let $G^t_{CN} = (V, E^t_{CN})$ denote an undirected contact network graph where an edge $e^t_{CN} = (u, v)$ between u and v exists if they had at least one contact during $[t_s, t]$, where $t \in [t_s, t_e]$. Let $G^t_{Fused} = (V, E^t_{Fused})$ denote an undirected contact network graph where $E^t_{Fused} = E^t_{CN} \cup E_{OSN}$. The second problem is to infer edges in E_{CN} using the information about edges in E^t_{Fused}.

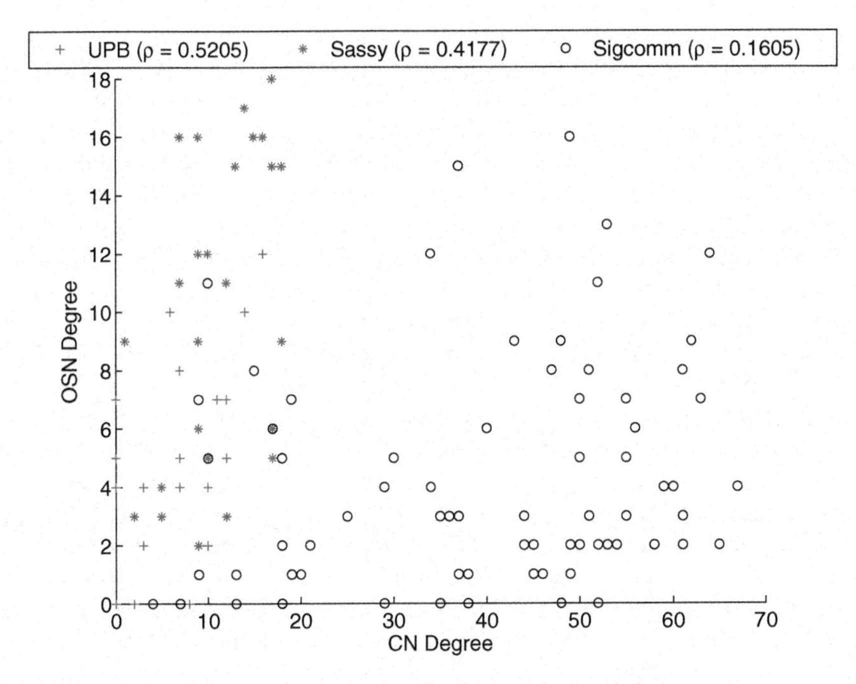

Fig. 4 Correlation between contact network and online social network degrees

5 Contact Graph Inference Using Online Social Information

As previously shown, there is a dependency between nodes' similarities and their levels of interaction. The following steps describe our algorithm for inferring edges in E_{CN} by selecting one of the previously adopted similarity measures.

Step 1: Store all the observed edges between node pairs in G_{CN} in a list l_e and sort this list in a decreasing order based on total number of contacts for each pair.

Step 2: Compute the similarity values between node pairs and store these node pairs in a list l_s sorted in a decreasing order based on similarity value for each pair.

Step 3: Extract those node pairs whose similarity values stored in l_s are greater than or equal to a similarity threshold th_s and store them in a list l_{th}.

Step 4: Store the number of pairs contained in l_{th} in $size_{lth}$.

Step 5: Extract the first $size_{lth}$ pairs from l_e and store them in a list l'_e.

Step 6: Store the number of matched pairs (correct predictions) between l_{th} and l'_e in mp_{th}.

In order to evaluate the performance of the similarity measures used as predictors, we use different similarity threshold values. For each similarity threshold, we compute the percentage of true positives vs. the percentage of node pairs having a similarity greater than the considered threshold.

In this section, we assume that we have only information about individuals' online social network graph while their contacts are completely unknown. We, therefore, infer human contacts by computing their similarity on an online social network. To the best of our knowledge, this is the first work that addresses the problem of predicting the formation of a link in a human contact network by using the corresponding online social network. In Fig. 5, the prediction results shown for the three datasets demonstrate that using online social data without any mobility information is helpful for predicting the links of a contact graph. As can be seen, we compare the performance of the chosen similarity measures to a random predictor that simply guesses links at random. In UPB dataset, *resource allocation* performs the best except for large values of similarity thresholds (using a smaller percentage of the population) where its performance is similar to *Jaccard coefficient* and *Adamic Adar*. As we decrease the similarity threshold, the effect of online social similarity decreases leading to results similar to random predictor. Note that the values of true positives for small percentages of the population indicate the percentages of correctly predicted strong ties. In Sassy and Sigcomm, on the contrary, the performance of our predictors are more similar to a random predictor even for small percentages of populations. As previously shown, online behavior in Sassy, and especially in Sigcomm has an impact on contact patterns lower than in UPB. However, the overall percentage of matching between our predictors and the observed edges in the contact network graph is higher than a random predictor. Looking more closely at Sassy and Sigcomm results, in the former, *resource allocation* performs better than the other measures for most percentages of population; in the latter, all measures except for *preferential attachment* show similar performance.

6 Contact Graph Inference Using Partial Contact Information

In the previous section, we have shown that nodes that are more similar on online social networks have a tendency to contact more often. In this section, we want to study the usefulness of both online and offline nodes' similarity for contact prediction. Specifically, we consider the problem of predicting the contact graph using both the online social network and part of the contact network.

As a first step, we analyze the effectiveness of using the first day of contacts of each dataset for predicting the overall contact network graph. The results in Fig. 6 show clearly that the knowledge of part of the contact data gives good prediction results if compared to the previous results. As can be seen, the performance of our predictors across all datasets are much more significant than random predictor. In particular, while in Sigcomm all our predictors show a very similar performance, in UPB *preferential attachment* and *resource allocation* perform best; in Sassy, *preferential attachment* and *Adamic Adar* are able to predict contact network graph better than other predictors.

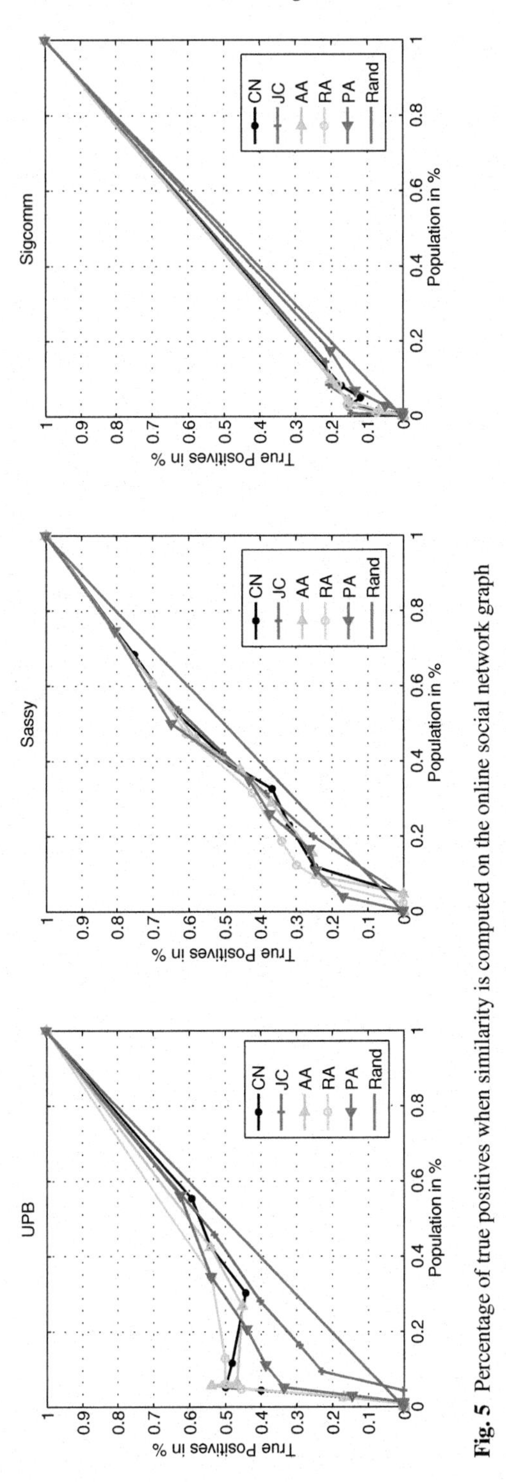

Fig. 5 Percentage of true positives when similarity is computed on the online social network graph

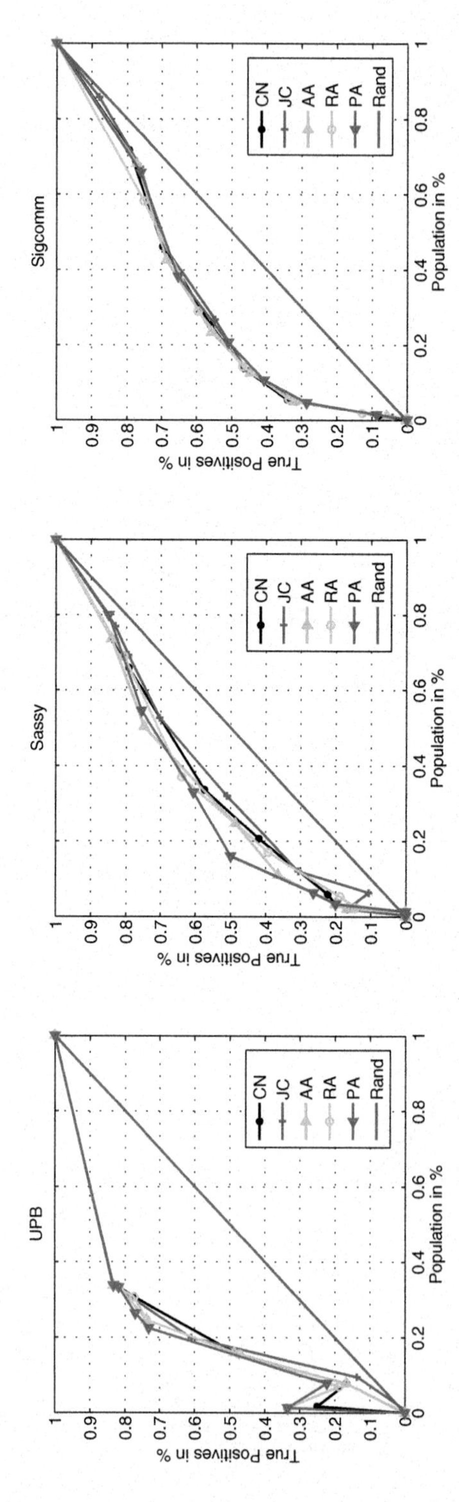

Fig. 6 Percentage of true positives when similarity is computed on the contact network graph representing the first day of contact data

If we combine the online social network graph with the partial contact graph, we obtain a fused network graph which reflects both human and online interactions between users. Using this graph for computing similarity between nodes, we obtain the results shown in Fig. 7. Analyzing the UPB dataset, we find an interesting result: adding online social information to partial contact network graph, predictors are able to better identify stronger ties. Note that for high similarity values (percentages of population lower than 0.1) all predictors perform significantly better by producing a higher percentage of true positives. We can further observe that *resource allocation* performs best in most cases.

Online social information has the same effect in Sassy and Sigcomm. For high similarity values, it is able to improve the performance of predictors. Considering the prediction results of each similarity measure, *resource allocation*, *preferential attachment* and *Adamic Adar* are good predictors for Sassy, while in UPB, *resource allocation* performs slightly better than the other measures.

7 Conclusion

In this paper, we studied the novel problem of predicting contacts using online social information. Motivated by the homophily theory, we selected a set of similarity measures and tested their predictive power on three different mobility traces. Formulating the problem of human contact prediction as a graph inference problem, we showed that it is possible to predict the edges of a contact graph when only the information about the links in the online social network graph is available.

Furthermore, we studied a similar problem in a different setting. We predicted the contact graph using both the online social network and part of the contact network, showing that online information is able to improve the predictive power of similarity measures computed on the contact graph by identifying stronger ties. In both settings, we showed that our proposed predictors outperform a random predictor and *resource allocation* plays a significant role among all datasets.

The problem of inferring human contacts using online social ties represents a novel and interesting research direction in social network analysis. There are many potential future directions of this work. First, it is interesting to combine several similarity features in order to study if this could enhance the prediction results. Next, other social theories can be explored for analyzing the predictive power of online social information. Another potential issue is to validate the proposed link predictors on some other mobility traces.

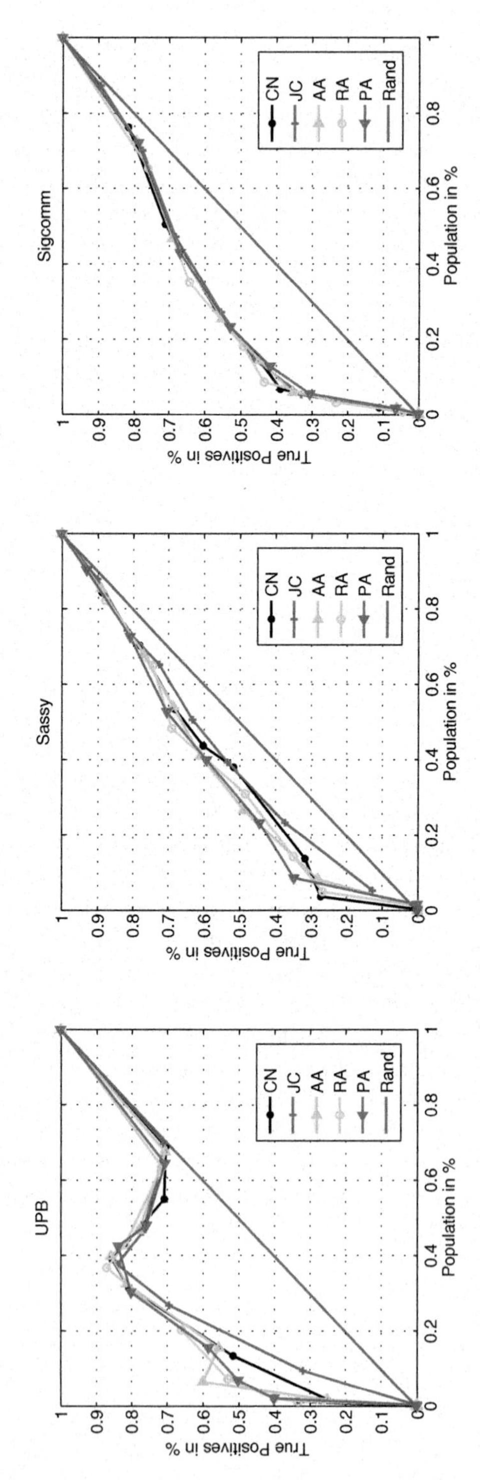

Fig. 7 Percentage of true positives when online social network graph and partial contact network graph are fused, and similarity is computed on this graph

References

1. Adamic, L.A., Adar, E.: Friends and neighbors on the web. Soc. Netw. **25**(3), 211–230 (2003)
2. Barabási A.-L.: Linked: the new science of networks. Perseus, New York (2002)
3. Bigwood, G., Rehunathan, D., Bateman, M., Henderson, T., Bhatti, S.: Exploiting self-reported social networks for routing in ubiquitous computing environments. In: networking and communications, 2008. WIMOB '08. IEEE international conference on wireless and mobile computing, pp. 484–489, Oct 2008
4. Bigwood, G., Rehunathan, D., Bateman, M., Henderson, T., Bhatti, S.: CRAWDAD data set st_andrews/sassy (v. 2011-06-03). http://crawdad.cs.dartmouth.edu/st_andrews/sassy (2011). Accessed June 2011
5. Ciobanu, R.I., Dobre, C.: CRAWDAD data set upb/mobility2011 (v. 2012-06-18). http://crawdad.cs.dartmouth.edu/upb/mobility2011 (2012). Accessed June 2012
6. Ciobanu, R.I., Dobre, C., Cristea, V.: Social aspects to support opportunistic networks in an academic environment. In: Ad-hoc, Mobile, and Wireless Networks, pp. 69–82. Springer, Berlin (2012)
7. Hsu, W.-J., Spyropoulos, T., Psounis, K., Helmy, A.: Modeling spatial and temporal dependencies of user mobility in wireless mobile networks. IEEE/ACM. Trans. Netw. **17**(5), 1564–1577 (2009)
8. Hui, P., Chaintreau, A., Scott, J., Gass, R., Crowcroft, J., Diot, C.: Pocket switched networks and human mobility in conference environments. Proceedings of the 2005 ACM SIGCOMM workshop on delay-tolerant networking, WDTN '05, pp. 244–251. ACM, New York (2005)
9. Jahanbakhsh, K., Shoja, G.C., King, V.: Human contact prediction using contact graph inference. Proceedings of the 2010 IEEE/ACM int'l conference on green computing and communications & int'l conference on cyber, physical and social computing, GREENCOM-CPSCOM '10, pp. 813–818. IEEE Computer Society, Washington DC (2010)
10. Jahanbakhsh, K., King, V., Shoja, G.C.: Predicting missing contacts in mobile social networks. In: World of Wireless, Mobile and Multimedia Networks (WoWMoM), 2011 IEEE international symposium on a, pp. 1–9, 2011
11. Jahanbakhsh, K., King, V., Shoja, G.C.: Predicting human contacts in mobile social networks using supervised learning. Proceedings of the fourth annual workshop on simplifying complex networks for practitioners, SIMPLEX '12, pp. 37–42. ACM, New York (2012)
12. Kevin, F.: A delay-tolerant network architecture for challenged internets. Proceedings of the 2003 conference on applications, technologies, architectures, and protocols for computer communications, SIGCOMM '03, pp. 27–34. ACM, New York (2003)
13. Kostakos, V., Venkatanathan, J.: Making friends in life and online: equivalence, micro-correlation and value in spatial and transpatial social networks. Proceedings of the 2010 IEEE second international conference on social computing, SOCIALCOM '10, pp. 587–594. IEEE Computer Society, Washington, DC (2010)
14. Liben-Nowell, D., Kleinberg, J.: The link-prediction problem for social networks. J. Am. Soc. Inf. Sci. Technol. **58**(7), 1019–1031 (2007)
15. McPherson, M., Smith-Lovin, L., Cook, J.M.: Birds of a feather: homophily in social networks. Annu. Rev. Sociol. **27**, 415–444 (2001)
16. Mirco, M., Cecilia, M.: A community based mobility model for ad hoc network research. Proceedings of the 2nd international workshop on multi-hop ad hoc networks: from theory to reality, pp. 31–38. ACM, New York (2006)
17. Mtibaa, A., May, M., Ammar, M.: On the relevance of social information to opportunistic forwarding. In: Modeling, Analysis & Simulation of Computer and Telecommunication Systems (MASCOTS), 2010 IEEE International Symposium on, pp. 141–150. IEEE, Washington DC (2010)
18. Pan, S.J., Boston, D.J., Borcea, C.: Analysis of fusing online and co-presence social networks. In: Pervasive Computing and Communications Workshops (PERCOM Workshops), 2011 IEEE international conference on, pp. 496–501, March 2011

19. Pietiläinen, A.-K., Diot, C.: Dissemination in opportunistic social networks: the role of temporal communities. Proceedings of the thirteenth ACM international symposium on mobile ad hoc networking and computing, MobiHoc '12, pp. 165–174. ACM, New York (2012)
20. Pietilainen, A.-K., Diot, C.: CRAWDAD data set thlab/sigcomm2009 (v. 2012-07-15). http://crawdad.cs.dartmouth.edu/thlab/sigcomm2009 (2012). Accessed July 2012
21. Scholz, C., Atzmueller, M., Stumme, G.: On the predictability of human contacts: influence factors and the strength of stronger ties. In: Privacy, Security, Risk and Trust (PASSAT), 2012 international conference on Social Computing (SocialCom), pp. 312–321, 2012
22. Socievole, A., Marano, S.: Exploring user sociocentric and egocentric behaviors in online and detected social networks. In: Future Internet Communications (BCFIC), 2012 2nd Baltic Congress on, pp. 140–147, April 2012
23. Song, C., Qu, Z., Blumm, N., Barabási, A.-L.: Limits of predictability in human mobility. Science. **327**(5968), 1018–1021 (2010)
24. Vu, L., Do, Q., Nahrstedt, K.: Jyotish: constructive approach for context predictions of people movement from joint wifi/bluetooth trace. Pervasive Mob. Comput. **7**(6), 690–704 (2011)
25. Wang, D., Pedreschi, D., Song, C., Giannotti, F., Barabasi, A.-L.: Human mobility, social ties, and link prediction. Proceedings of the 17th ACM SIGKDD international conference on Knowledge Discovery and Data Mining, KDD '11, pp. 1100–1108. ACM, New York (2011)
26. Zhou, T., Lü, L., Zhang, Y.-C.: Predicting missing links via local information. Eur. Phys. J. B. **71**(4), 623–630 (2009)

Identifying a Typology of Players Based on Longitudinal Game Data

Iftekhar Ahmed, Amogh Mahapatra, Marshall Scott Poole, Jaideep Srivastava and Channing Brown

Abstract This study describes an approach to identify a typology of players based on longitudinal game data. The study explored anonymous user log data of 1854 players of EverQuest II (EQII)—a massively multiplayer online game (MMOG). The study tracked ten specific in-game player behavior including types of activities, activity related rewards, and casualties for 27 weeks. The objective of the study was to understand player characteristics and behavior from longitudinal data. Primary analysis revealed meaningful typologies, differences among players based on identified typologies, and differences between individual and group related gaming situations.

1 Introduction

We are currently experiencing a rapid growth of MMOGs and the number of players playing these games. The majority of the massively multiplayer online games (MMOGs) allows task-related activities and also provides a rich context for social experience. It is important to understand these games and players as MMOG users may have different perspectives on task or social life and may prefer spending their social time and energy in game environments rather than in the real world [1]). Moreover, for different people there could be different motivations as well as different meanings and consequences of playing the same game [2].

Current research has already provided insight into community building aspects of online games [3] behavioral analysis based on virtual world of games, game-based learning experiences, and group formation [4]. By providing tailored learning environments, a few specially designed games also allow learners to explore situations

I. Ahmed (✉)
University of North Texas, Denton, TX, USA
e-mail: Iftekhar.ahmed@unt.edu

A. Mahapatra · J. Srivastava
University of Minnesota, Union Street SE 200, Keller 5-209,
Minneapolis, MN 55455, USA
e-mail: srivasta@cs.umn.edu

M. S. Poole · C. Brown
University of Illinois at Urbana Champaign, Urbana, Illinois

M. A. Ahmad et al. (eds.), *Predicting Real World Behaviors from Virtual World Data*,
Springer Proceedings in Complexity, DOI 10.1007/978-3-319-07142-8_7,

that are either unsafe (i.e., the battlefield) or costly (i.e., flight simulation) to create in the real world [5]. A significant number of studies related to socio-psychological aspects (motivation, community building, social networking, etc.) of players and technical issues (net traffic, visualization and graphics, control mechanism, etc.) exist. However, few of these studies have focused on MMOGs over time. Keeping track of individual actions and behaviors adds considerable challenge in the data collection and analysis process. Limitations in game studies also exist due to the proprietary nature of the industry and maintaining player privacy in research [6]. However, the importance of understanding players and their behavior over time cannot be ignored.

This research attempts to identify a typology of players in EverQuest II (EQII), a MMOG, by looking at their actions and behaviors over time. The research explored ten specific in-game aspects over a 27-week period to developed player-typologies. The objective of this project is to understand player characteristics and behavior from their activities over time. This is an ongoing research project and will in the future explore relationships between the typology and players' socio-psychological characteristics. However, the primary analysis reported in the chapter already reveals significant insight into player categories.

2 Data

This research explored anonymized user log data of EQII players. The data set contains 1854 players' in-game activities over 27 weeks' time period. It includes types of activities, activity-related rewards, and casualties. EQII gaming activities are designed around quests. A quest refers to a specific in-game adventure or endeavor. From a grouping perspective, EQII allows two different types of quests—solo and group. Solo quests are designed for single player and group quests are designed for multiplayer activities. There are in-game rewards and punishments associated with quests. For this research, we considered three different reward types—achievement and experience points earned by the players, and leveling up. In-game death by the hand of artificial opponents (such as monsters) was considered as casualties. Experience points and deaths were linked to either an individual- or group-related task. Total number of individual (solo) and group quests played by a player, achievement points earned for those quests, experience points earned for both solo and group quests, deaths occurred in solo and group quests, level of opposition in quests, and changes in player level over time were the study variables. These variables were chosen as they capture the very basic nature of the game-play based on game mechanics. Original data logs contained time-stamped data for each action and consequences. Weekly counts of all these variables for each individual were calculated for analysis.

3 Method

We employed cluster analysis to identify player typologies. In the absence of prior theoretical reasoning to decide on cluster numbers, we first utilized a hierarchical clustering method (Wards method with Squared Euclidian Distance measure) as

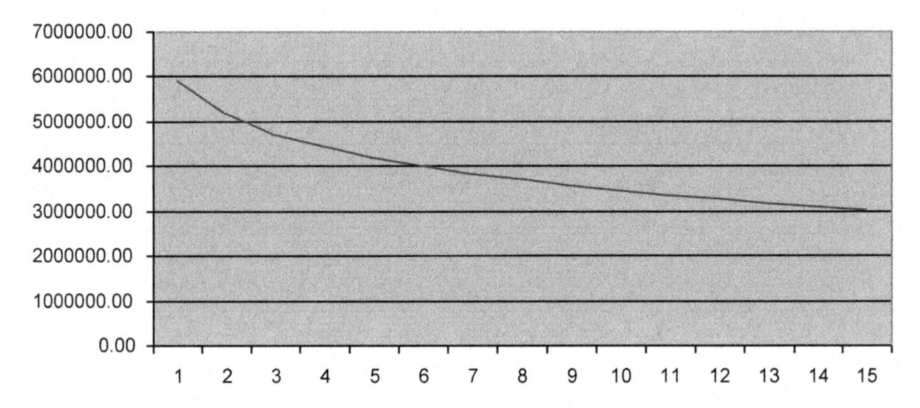

Fig. 1 Elbow showing solo quests over time clustering

exploratory cluster analysis. One of the recognized ways to decide on the optimal cluster number is to plot the number of clusters against the distance at which clusters are combined and looking for an elbow after plotting [7]. We employed this method using a composite measure of all variables and obtained cluster solutions.

Once we got the optimal cluster number, we applied document clustering techniques for clustering [8]. In the data, each player was represented by a 27-attribute-long feature vector. Each attribute captured a certain characteristic, e.g., number of deaths, in a certain week. This is a high-dimensional feature space and hence we used the state-of-the-art document clustering techniques which have been shown to work well in high dimensions. We used the repeated bisection algorithm with cosine as our distance measure to carry out the clustering process. Cosine is known to work well in high dimensions as compared to Euclidean distance.

Finally, for each variable, we generated weekly mean values for each clustered group to identify their nature and differences among clustered categories.

4 Results

This section details player categories based on the analysis variables. The section, first, reports the findings of the optimal cluster number and then describes the difference between/among clustered categories.

5 Differences in Quest-Play

The initial screen plot elbow indicates that there are three potential categories of players based on the number of solo quests they played over time (Fig. 1). Figure 2 illustrates cluster means of solo quest-play over the sampling period. Cluster 1 players

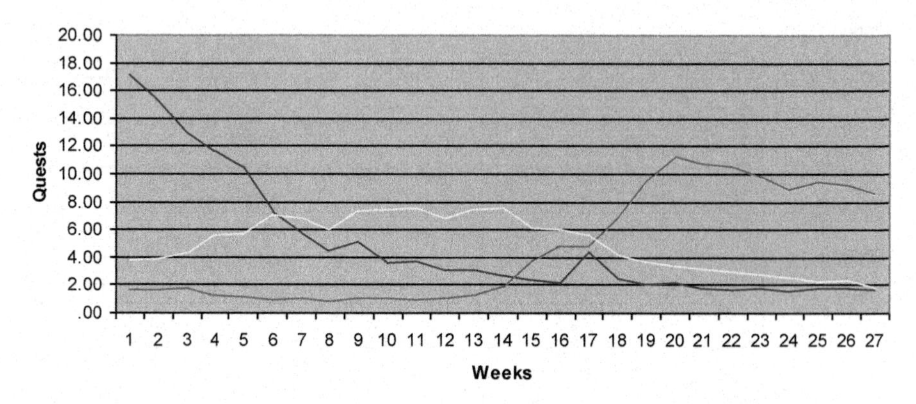

Fig. 2 Cluster means of solo quests

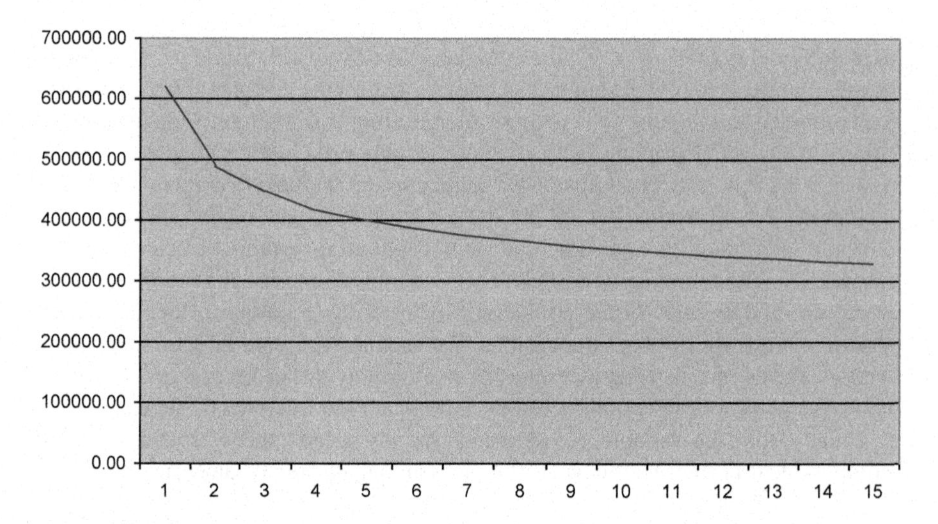

Fig. 3 Elbow showing group quests over time clustering

showed a higher number of quest play in the initial week ($M = 17.18$) but then a gradual decline over time ($M = 1.64$). Cluster 2 players showed initial low number of quests ($M = 3.82$) that further declines over time ($M = 1.82$). Opposite to cluster 1, cluster 3 player started with a low number of quests ($M = 1.64$), and subsequently increased the number of quest play over time ($M = 8.56$).

Initial screen plot of group quest-play indicated a 4-cluster solution (Fig. 3). Figure 4 illustrates cluster means of group quest-play. Similar to solo quest, group quest cluster 1 indicated a higher number of quests in the initial week ($M = 6.59$) which then slowly declined over time ($M = 0.79$), and cluster 2 players showed initial low number of quests ($M = 1.90$) that further declined over time (1.07). Interestingly, we observed a very different pattern of playing behavior in cluster 3. Players within this category started low ($M = 1.41$), increased the frequency of play after 11th

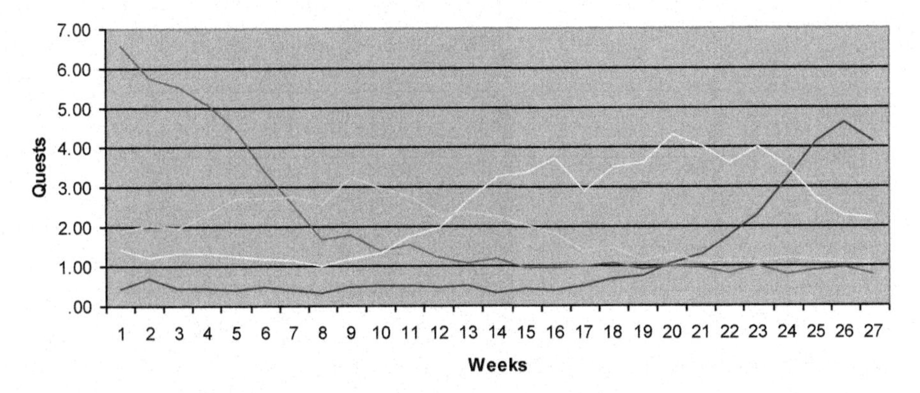

Fig. 4 Cluster means of group quests

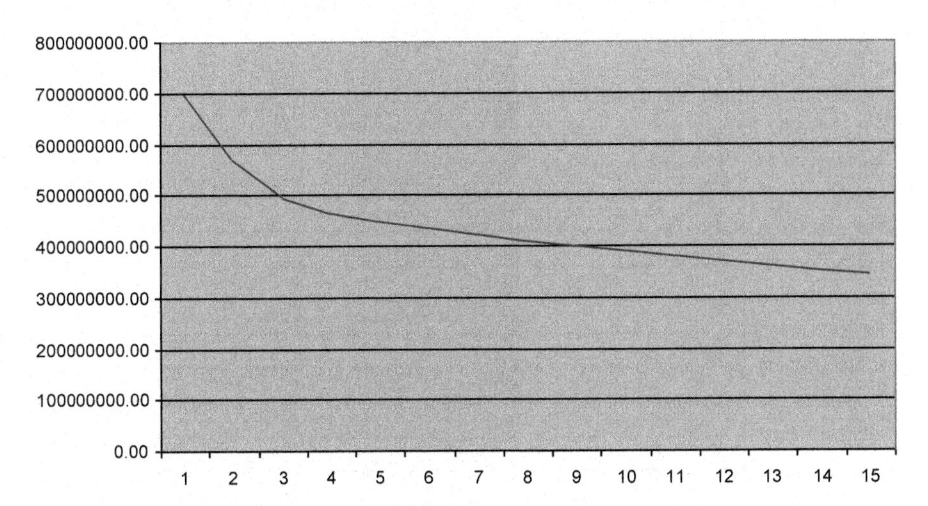

Fig. 5 Elbow showing achievement points over time clustering

week ($M = 4.34$), but then started showing a decrease in the number of game playing frequency ($M = 2.22$). Cluster 4, similar to solo game cluster 3, started as the lowest ($M = 0.42$) but gained momentum over time and ended as the highest playing cluster (4.22).

5.1 *Differences in Earning Rewards*

Initial screen plot elbow indicated three clusters for achievement points over 27 weeks (Fig. 5). Cluster 1 players started with the highest mean values (260.77) and ended with the second lowest mean values (21.21) in earning achievement points (Fig. 6). Cluster 2 players started with low moderate mean values (42.00), climbed

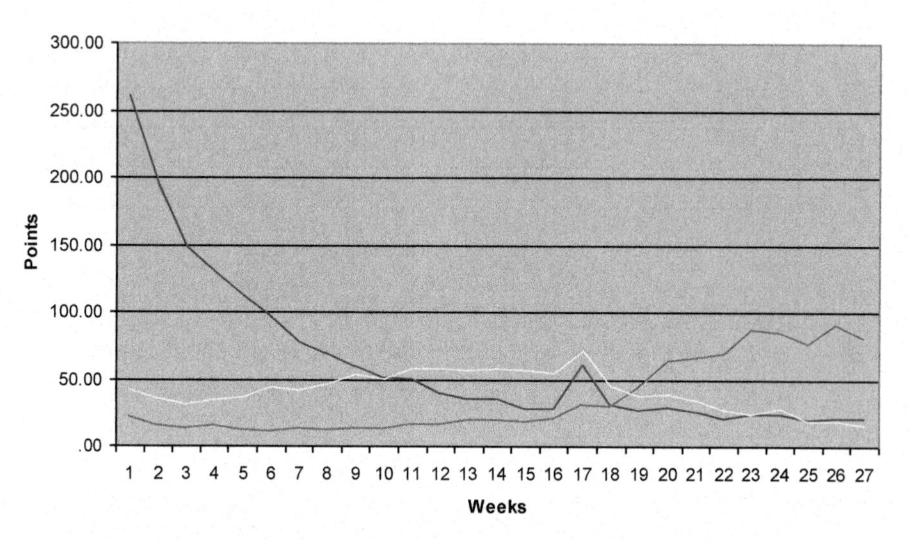

Fig. 6 Cluster means of achievement points

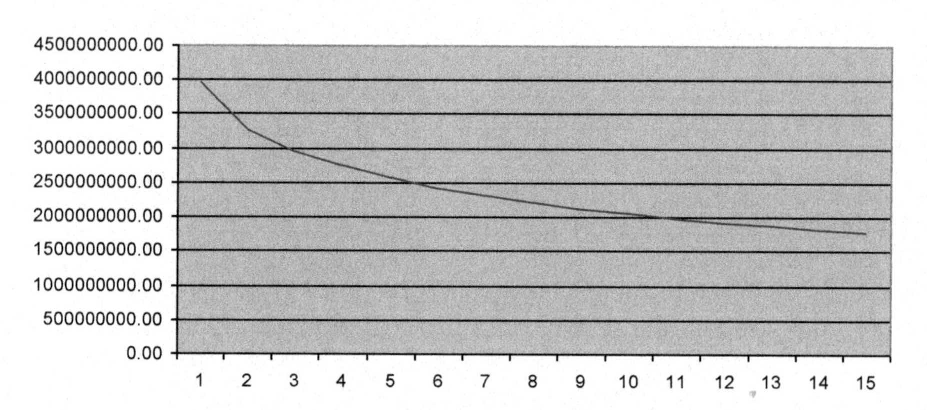

Fig. 7 Elbow showing experience points in solo quests over time clustering

to a moderate mean value (71.09), and finally slumped to the lowest mean value (16.00) among all groups. Cluster 3 players started with the lowest starting mean value (21.67) but ended with the highest end mean value (81.22).

Initial analysis of experience points while playing alone indicated a 2-cluster solution (Fig. 7). Analysis showed completely opposing trends for the two clusters (Fig. 8). Cluster 1 players, started with a very high experience point mean (263.90) but finished with the lowest recorded score (32.43). On the other hand, cluster 2 slowly rose to a moderately high score from an initial 59.91 to 156.17.

Earning experience points while playing with groups showed four distinctive clusters (Fig. 9). Cluster 1 started with a large value of 2,594 but slumped to 141.42, which is the lowest among all clusters (Fig. 10). Cluster 2 started with a moderate 624.75, gained momentum in the beginning (922.44) before slumping down to a low

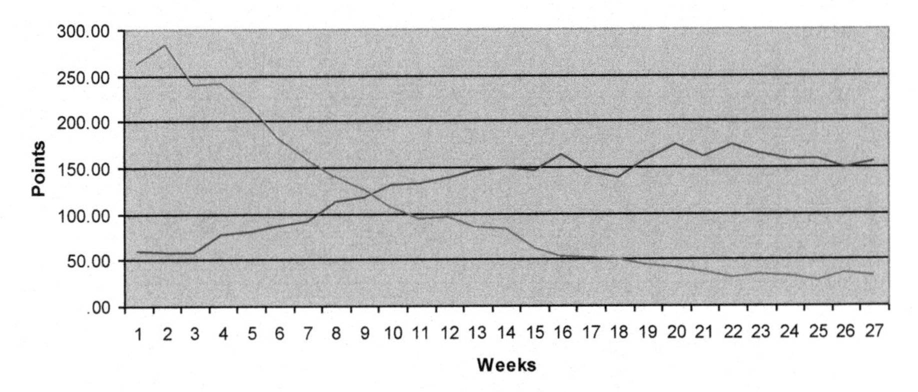

Fig. 8 Cluster means of experience points in solo quests

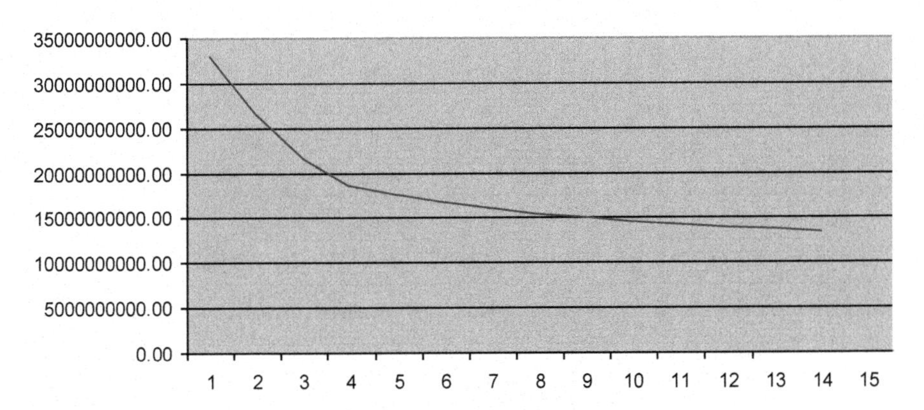

Fig. 9 Elbow showing experience points in group quest over time clustering

value (158.37). Cluster 3 started low (180.25) and gained a little momentum in the middle (513.41) before going down to low again (162.72). Cluster 4 started with the lowest among all values (74.97) but started rising after the 17th week before ending as the highest end value among all clusters (663.53).

One significant difference in experience points is that group-quest-related experience points are much higher than points earned by players playing individual quests. This difference is due to the game mechanics.

Players leveling up over time revealed two distinctive clusters (Fig. 11). Cluster 1 contained players with moderate player level. Players within this cluster have an initial mean value of 36.13 that stayed almost flat till the end (29.37). However, cluster 2 contained mostly new players (started with 4.61) who were slowly leveling up (ended with 22.34). These two clusters are so distinctive that for the whole duration of 27 weeks, cluster curves remained separated (Fig. 12).

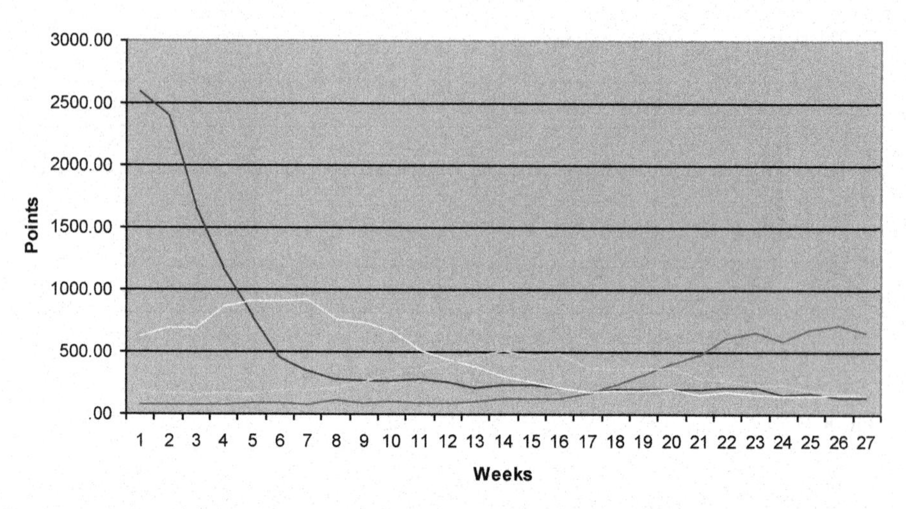

Fig. 10 Cluster means of experience points in group quests

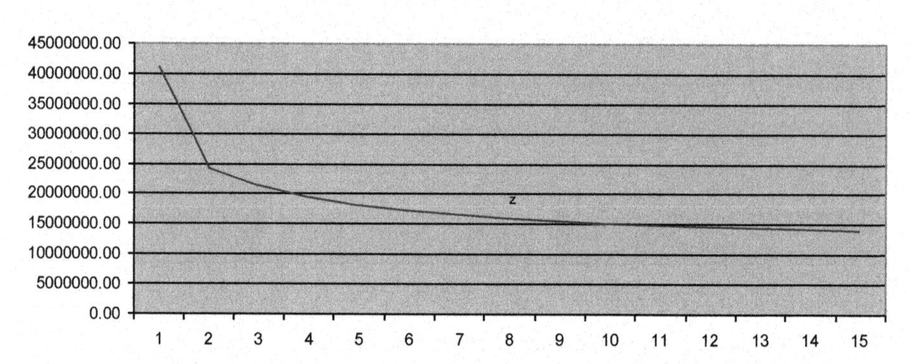

Fig. 11 Elbow showing leveling up over time clustering

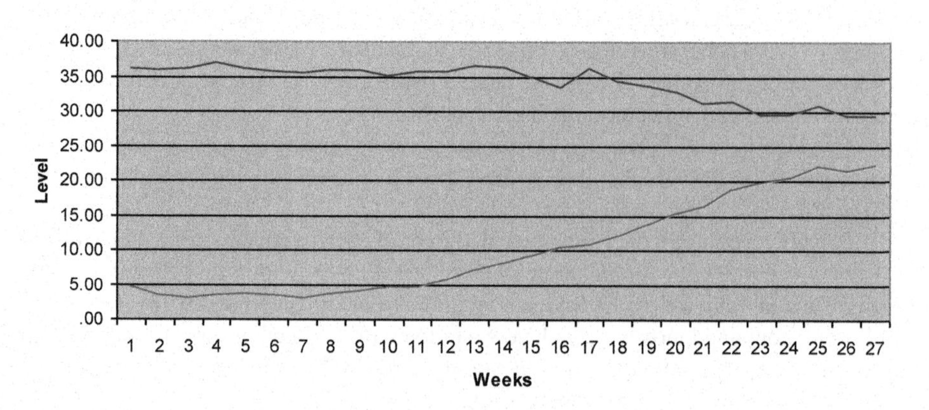

Fig. 12 Cluster means of leveling up

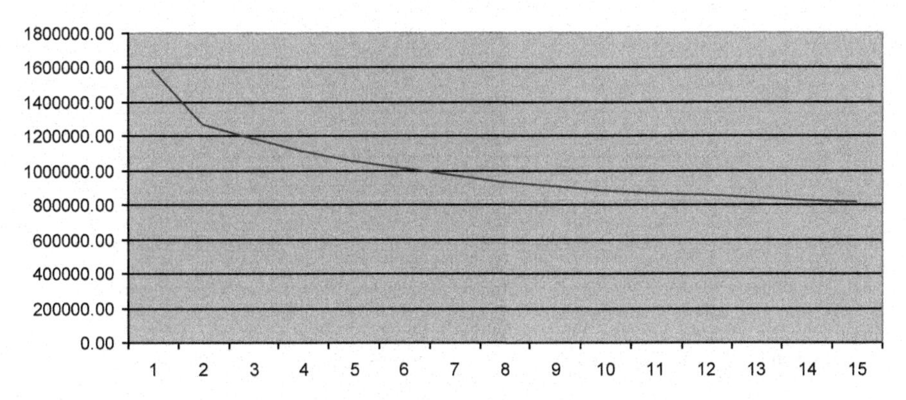

Fig. 13 Elbow showing deaths while playing alone over time clustering

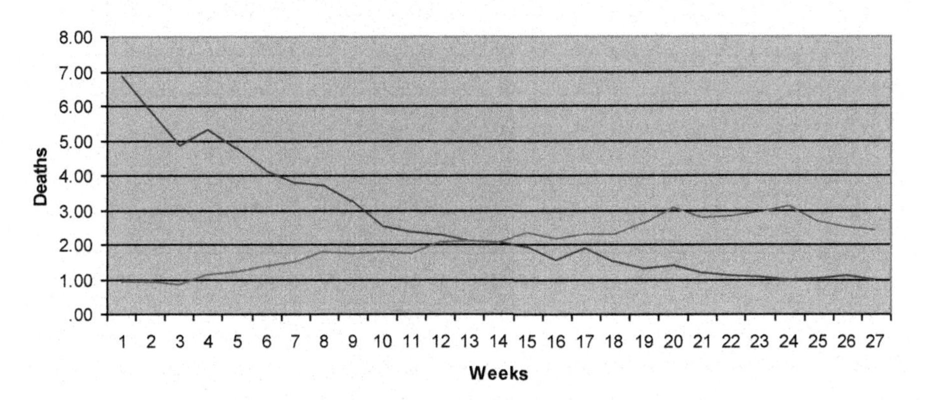

Fig. 14 Cluster means of deaths while playing alone

5.2 *Differences in Casualties*

Analysis of the number of deaths while playing alone revealed two clusters (Fig. 13). These clusters were very different from each other (Fig. 14). Cluster 1 players started with a high death rate (6.87) but that climbed down to a low value (0.97). Whereas, cluster 2 players started with low rates (0.96) but the rate slowly increased (2.40 at the end).

On the other hand, deaths while playing with groups revealed three clusters (Fig. 15). While two of the clusters are similar to those for solo deaths, we observed a different group who maintained a moderately low value throughout 27 weeks (Fig. 16). This group started with 4.23 average death counts and ended up with 4.09 average death counts (cluster 3). Two other clusters similar to the previous results are the high beginners who ended low (cluster 1) and low beginners who ended up high (cluster 2).

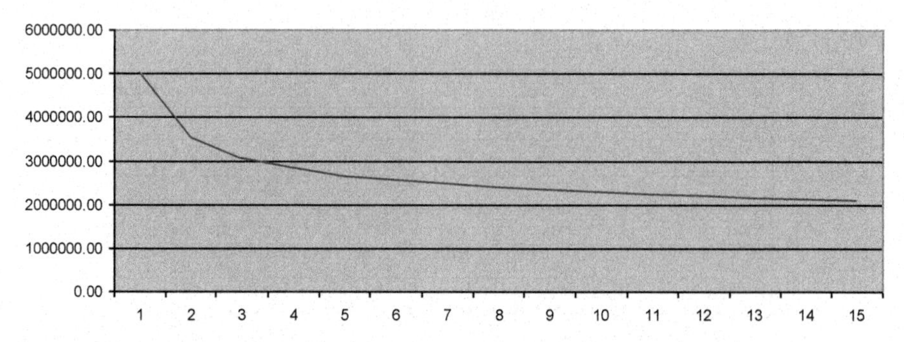

Fig. 15 Elbow showing deaths while playing with groups over time clustering

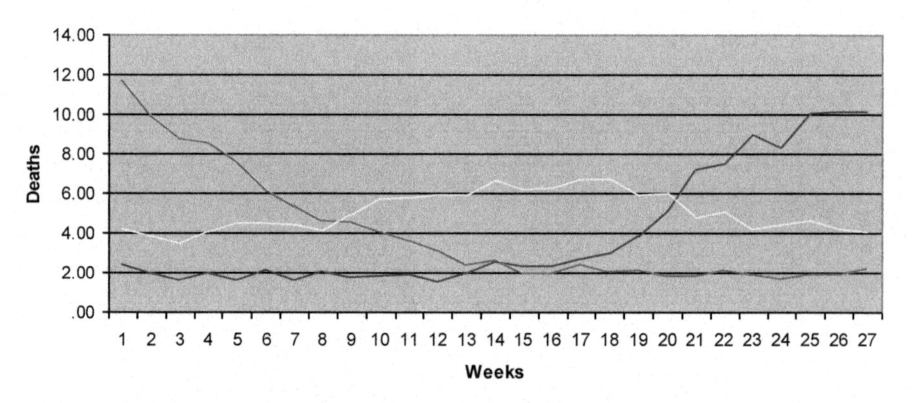

Fig. 16 Cluster means of deaths while playing with groups

5.3 *Differences in Opponent Level*

We observed three different clusters of players based on the level of their opponent when they play alone (Fig. 17). Cluster 1 players started with moderately high level (39.44) opponents but slowly went down to low level opponents (5.65) (Fig. 18). Opposite to that, cluster 3 players started with low level opponents (3.04) and rose to moderate opponents (15.14). Cluster 2 players, on the other hand, maintained an almost equilibrium in choosing their opponents (started with 26.68 and ended up with 20.94).

We identified two clusters among players playing group quests (Fig. 19). The clusters also vary in nature while comparing to solo games (Fig. 20). Cluster 1 players started with moderately high level opponents (34.67) and came down to moderate opponents (21.60) at the end. Whereas, cluster 2 players started with low level opponents (6.95) but ended almost around the same place as the first cluster (22.33). It is also important to mention that, in an average, opponent level in group quest remained higher than solo quest opponents.

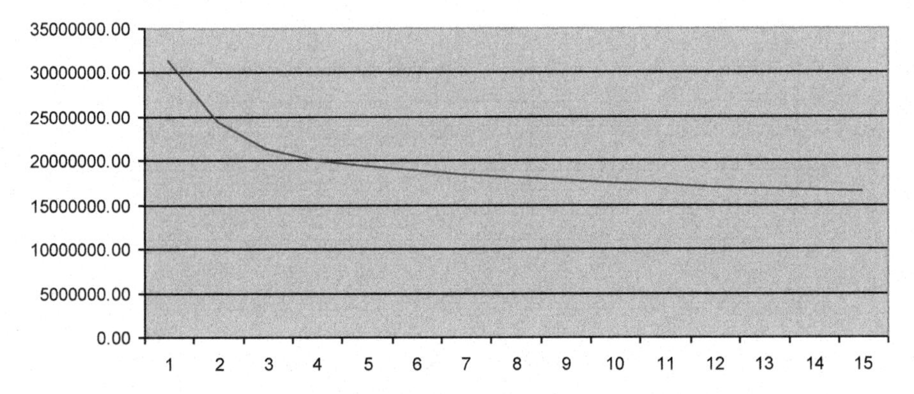

Fig. 17 Elbow showing opponent level while playing alone over time clustering

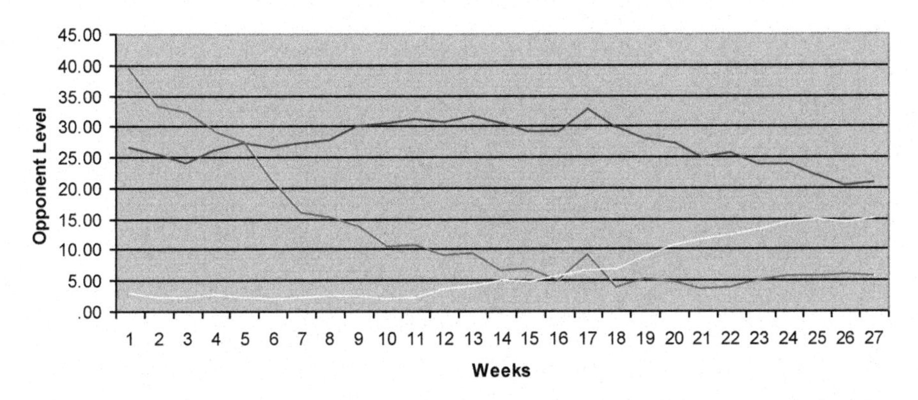

Fig. 18 Cluster means of opponents while playing alone

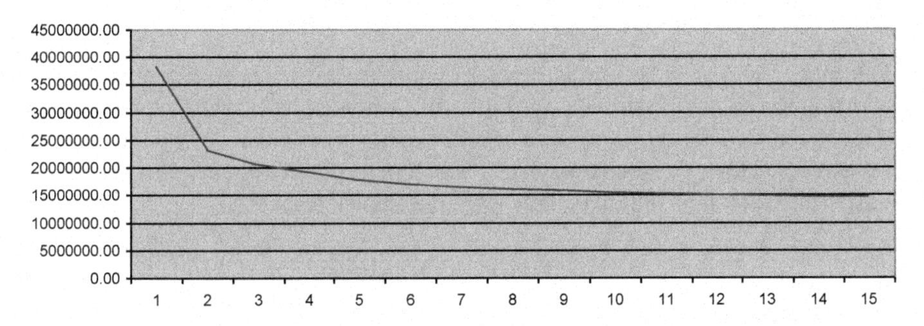

Fig. 19 Elbow showing opponent level while playing with group over time clustering

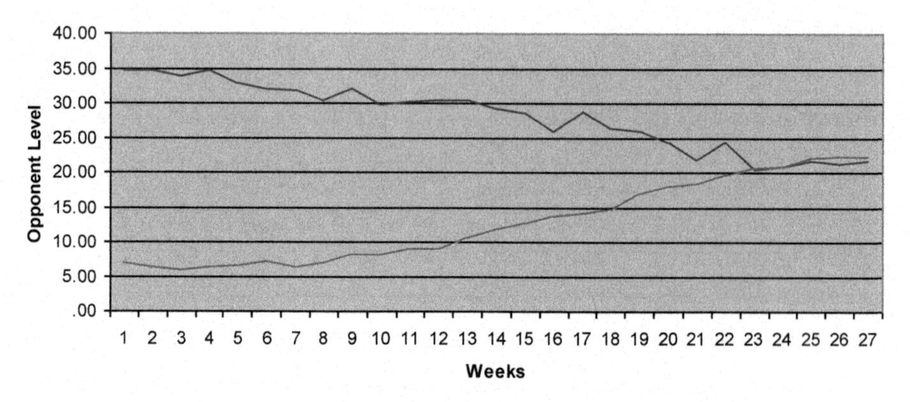

Fig. 20 Cluster means of opponent while playing with groups

Table 1 Distribution of players in clusters

Clustering variable	Clusters			
	1	2	3	4
Achievement points	430	979	445	–
Experience points (group)	349	392	769	344
Experience points (solo)	1,041	813	–	–
Deaths (group)	505	368	981	–
Deaths (solo)	788	1,066	–	–
Quests (group)	393	854	407	200
Quests (solo)	658	846	350	–
Opponent level (group)	924	930	–	–
Opponent level (solo)	363	803	–	–
Player level	1,174	680	688	–

5.4 *Distribution of Players Across Clusters*

Our analysis revealed that the distribution of players across clusters was not balanced and sometimes much skewed (see Table 1). Opponent level while playing group quests produced almost equal size clusters. Group quests cluster 4 was the smallest observed cluster with approximately 11 % and player level cluster 1 was the largest observed cluster with approximately 64 % of sampled players.

6 Discussion

Our analysis identified that there are meaningful categories among players based on different game playing aspects over time. First, we have observed that the player categories were distinctive within each aspect of play (i.e., deaths, points, etc.). Second, when we compared solo with group play, categories of similar variables also portrayed differences. However, there were two dominant trends among categories.

One category of players showed consistent decline in values and the other showed the opposite tendency. This particular trend indicates that player motivation fluctuates over time. Further analysis will take these trends into account to investigate if there is any relationship between trends and player's psychological orientation.

We have also observed that our data analysis approach worked well with this particular data set. The fact that document clustering approach worked well with mapping the activities of players online suggests that the activity values like death, points, or quests, must be following a word frequency distribution, which is essentially a power law distribution. It means that most people have a similar kind of performance and we also have a long tail of outliers. Another aspect worth noticing is that the cosine vector worked better than similarity measures such as correlation. It tells us that the presence of activity is more predictive than its absence.

Acknowledgments This research was supported by the National Science Foundation (NSF IIS-0729421), the Army Research Institute (ARI W91WAW-08-C-0106), Air Force Research Lab (AFRL Contract No. FA8650-10-C-7010), the Army Research Lab (ARL) Network Science—Collaborative Technology Alliance (NS-CTA) via BBN TECH/W911NF-09-2-0053, and the National Science Foundation via the XSEDE project's Extended Collaborative Support Service under Grant # NSF-OCI 1053575. The data used for this research was provided by the SONY Online Entertainment. The findings represent solely the opinions of the authors and not of the sponsors.

References

1. Ng, B., Wiemer-Hastings, P.: Addiction to the internet and online gaming. CyberPsychology Behav. **8**(2):110–113 (2005)
2. Yee, N.: Motivations of play in online games. CyberPsychology Behav. **9**, 772–775 (2007)
3. Ducheneaut, N., Moore, R.: The social side of gaming: A study of interaction patterns in a massively multiplayer online game. Proceedings of the CSCW' 04, pp. 360–369 (2004)
4. Ahmed, I., Brown, C., Pilny, A., Cai, D., Ada, Y. A., Poole, M. S.: Identification of groups in online environments: The twist and turns of grouping groups. In: Privacy, security, risk and trust (passat), 2011 IEEE third international conference on and 2011 IEEE third international conference on social computing (SocialCom), IEEE, pp. 629–632. (2011)
5. Susi, T., Johannesson, M., Backlund, P.: Serious games: An overview. Technical Report HS-IKI-TR-07 – 01: School of Humanities and Informatics, Sweden (2007)
6. Feng, W., Brandt, D., Saha, D.: A long-term study of a popular mmorpg, NetGames '07: Proceedings of the 6th ACM SIGCOMM workshop on Network and system support for games, pp. 19–24 (2007)
7. Milligan, G.W., Cooper, M.C.: An examination of procedures for determining the number of clusters in a data set. Psychometrika. **50**, 159–79 (1984)
8. Zhao, Y., Karypis, G.: Evaluation of hierarchical clustering algorithms for document datasets. 11th Conference of Information and Knowledge Management (CIKM), pp. 515–524 (2002)

Index

M. A. Ahmad et al. (eds.), *Predicting Real World Behaviors from Virtual World Data*, Springer Proceedings in Complexity, DOI 10.1007/978-3-319-07142-8,